Using your TI-Nspire in learning electrical circuits

A REFERENCE TOOL BOOK FOR ELECTRICAL AND COMPUTER ENGINEERING STUDENTS AND PRACTITIONERS

Part I: Linear Resistive Circuits.

Rogelio Palomera-García
U. of Puerto Rico at Mayagüez

Using your TI-Nspire CX in learning electrical circuits
Part I: Linear Resistive Circuits.
Reference book for electrical and computer engineering students and practitioners

Copyright © 2016 by Rogelio Palomera-García.
All rights reserved, including the right to reproduce this book or any portion of it in any form.
   TI-Nspire, TI-Nspire CX, TI-Nspire CX CAS are trademarks of Texas Instruments.

Written and designed by:
  Rogelio Palomera-García
  r.palomera-garcia@ieee.org

ISBN-13: 978-1541118300
ISBN-10: 1541118308

   Printed by CreateSpace, an Amazon.com Company

Preface

> *The calculator is a tool, not a teacher. If you don't know how to do it by hand, with pencil and paper, then you don't know how to do it with the calculator.*

Following suggestions about the convenience of extending the book that I published, "Using your TI-89 in learning electrical circuits ", to other calculator models, I decided to publish this one and plan for others.

This is neither a textbook on circuit analysis nor a manual of your calculator nor a collection of programs. It is a reference book where you can learn to make an effective use of your graphing calculator as an aid to learn and apply circuits. The target audience for this book is composed mainly of electrical and computer engineering students, as well as practitioner engineers. However, non ece engineering and engineering technology students can also take advantage of it.

The focus of the book is in providing tools to help you in the process of learning. Programming is not the main contents of the book, although some programs illustrate the usefulness of the learned processes. You will find out soon that you can program easily for many applications.

One reason to work only with resistive circuits is to keep the length of the book within reasonable limits. Another one is to concentrate our efforts on the process and the use of the tool. Algebraically speaking, almost all the techniques presented may be extended to circuits in the phasor domain using complex numbers. Furthermore, many techniques in time domain of circuits with reactive elements can be associated to resistive network analysis. Two examples in the last chapter illustrate these remarks.

Anyway, I hope to produce a second part to deal with linear RLC circuits in the time domain and complex phasor domain as well. The phasor domain has many more interesting applications than just transforming real numbers into complex numbers for the analyses. Hence, to include everything in just one volume is an overkill. Both for the reader and for me, the author. ;)

Let me now describe briefly the chapters. You are not obliged to read them in sequence and you may focus in some of them while skipping others.

Chapter 1 is a brief overview of views I learned from my professors about what circuit analysis is. My thoughts sketched in this chapter are reflected in the book organization and layout.

Chapter 2 provides an overview of some features of you calculator, while *chapter 3* reviews topics of interest for lists and matrices.

Chapter 4 introduces the theorems of source substitution, homogeneity and superposition. Your calculator becomes a powerful tool when you apply these theorems, as seen later.

Chapter 5 is a tour through reduction and transformation methods used in circuit analysis. Series-parallel transformations, delta and wye structures, voltage dividers and others are introduced here. One goal of the chapter is to stress planning to obtain real benefit of our tool. Another, to show clearly how the features of your calculator may be exploited when these methods are used. Several user defined functions are introduced in this chapter.

Chapters 6 and 7 work the traditional nodal and loop analyses, respectively. However, they also include the so called modified nodal and loop analyses, because . Programming for nodal analysis and the particular case of mesh analysis are introduced in these chapters, with examples. The chapter on loop analysis is shorter than that of nodal analysis because node based description is more direct and based only on the topology of the circuit. In fact, nodal analysis is the basis for the popular SPICE simulator.

Chapter 8 to *Chapter 10* illustrate the power of the tool beyond writing and solving equations, when you combine theory and algorithms. Let me say that these chapters are the real motivation of this work. Some colleagues saw me applying the calculator to circuits and encouraged me to write a book. These chapters cover transfer functions, Thevenin and Norton representations, and two-ports.

Chapter 11 is an appetizer with a short tour beyond the scope of this book, applying what has been learned in other situations. It may also serve as a motivation for the companion books. For me, to finish writing them. For you, to read them if they are published.

The techniques presented in the book are useful for technology assisted analysis. Let it be with Matlab^{TM}, Mathematica^{TM}, Mathcad^{TM}, spreadsheets, and so on. Even without a calculator, you may find this book useful.

I thank in advance informing about any error, grammar or technical, that you find. I know that errors are somehow analog to lies: if you repeat them constantly, you finish believing they are correct. So I may not see them anymore! You may contact me to leave your comments, criticism or say hello sending a message to r.palomera-garcia@ieee.org

I hope you enjoy the book and find it useful!

Rogelo Palomera-Garcia
r.palomera-garcia@ieee.org

Contents

Preface · i

1 Preliminary considerations · 1
  1.1 Physical vs. Model Circuits · 1
  1.2 Our focus: linear resistive circuits · 2
    1.2.1 Using this book for AC circuits · 3
    1.2.2 Circuit equations and solving methods · 3
    1.2.3 Hand analysis vs analysis with calculator · 5
    1.2.4 Programming and using programs · 6

2 Practical matters. · 7
  2.1 Texas Instruments calculators used in this book · 7
  2.2 Keys and key typing · 7
    2.2.1 Cursor keys · 8
    2.2.2 Key notation and key values · 9
    2.2.3 Remarks about menu keys · 9
    2.2.4 Grouping symbols and entry fields · 9
    2.2.5 Powers of 10 using EE · 10
    2.2.6 Notation for entry and its result · 10
  2.3 Scratchpad and documents · 10
  2.4 Settings for display and data input · 11
    2.4.1 Numeric Display Formats: · 12
        "Display Digits" menu · 12
        Exponential Format menu · 12
        Calculation menu · 13
    2.4.2 Angle mode · 14
    2.4.3 Real or Complex menu · 14
  2.5 Variables and Storage · 14
    2.5.1 The [var] key · 15
    2.5.2 Browsing the stack · 16
    2.5.3 Variable ANS · 16

		2.6	Using the colon (:) in entries	17
		2.7	User Defined Functions I	18
			2.7.1 Algebraic non-parameter functions	19
			2.7.2 Functions with parameter	20
		2.8	Programming	20
		2.9	Closing remarks	23

3 Lists and Matrices   24

- 3.1 Lists . . . 24
    - 3.1.1 Working with Lists . . . 25
    - 3.1.2 Lists and your calculator . . . 27
        - List variables . . . 27
        - Retrieving and displaying elements of a list: . . . 27
        - Editing the list . . . 27
        - Augmenting the list . . . 27
    - 3.1.3 List functions and list menus . . . 27
- 3.2 Matrices . . . 28
- 3.3 Basic definitions and operations . . . 28
    - 3.3.1 Some basic definitions . . . 29
    - 3.3.2 Submatrices . . . 29
    - 3.3.3 Partitioning of matrices . . . 30
    - 3.3.4 One remark on multiplication of matrices . . . 31
    - 3.3.5 Matrices and linear combinations . . . 31
- 3.4 Matrices and the TI-nspire cx . . . 32
    - 3.4.1 Input of operations . . . 32
        - Dot operations in the TI-nspire cx . . . 33
    - 3.4.2 Creating matrices . . . 33
        - Using the calculator editor tool . . . 33
        - On the command line . . . 33
    - 3.4.3 Retrieving and editing elements . . . 34
    - 3.4.4 Retrieving and editing rows and columns . . . 34
    - 3.4.5 Submatrices . . . 35
    - 3.4.6 Matrix functions in the matrix menu . . . 36
        - Using `augment()` function . . . 37
- 3.5 Matrices and Linear equations . . . 38
    - 3.5.1 Solving linear equations . . . 39
    - 3.5.2 Multiple systems and knowns . . . 40
    - 3.5.3 Exchanging knowns and unknowns . . . 41
    - 3.5.4 Closing remarks . . . 43

4 Four Network Theorems and Applications   44

- 4.1 Substitution Theorems . . . 44
- 4.2 Homogeneity and Proportionality . . . 47
- 4.3 Superposition Theorem . . . 48
    - Writing equations for superposition . . . 51

		4.4	Closing remarks	52

5 Transformations and reductions — 53
- 5.1 Series and Parallel Connections 53
  - 5.1.1 Equivalent resistance and conductance formulas 54
  - 5.1.2 Another example of Series-Parallel Reduction 56
  - 5.1.3 Using Lists in series-parallel 58
  - 5.1.4 Function Parallel pl(z) 60
    - An example with Homogeneity property 64
- 5.2 Delta-Wye, Wye-Delta transformation 65
  - 5.2.1 Working formulas directly 66
  - 5.2.2 Using functions 67
- 5.3 Voltage and Current Dividers 70
  - 5.3.1 Voltage dividers again: using conductances 73
    - Loaded Divider 74
    - Passive adder with voltage divider: applying superposition 75
  - 5.3.2 Thevenin equivalent for two-resistors divider. 80
- 5.4 Source transformations 82
  - 5.4.1 Shifting theorems, and extended source transformations 84
- 5.5 Some operational amplifier structures 86
  - 5.5.1 A function for the OA structures 88
- 5.6 Closing Remarks 89

6 Nodal Analysis — 91
- 6.1 Introduction to nodal analysis 91
- 6.2 Only resistances and current sources 93
  - 6.2.1 Theoretical principles 94
    - Component of G_{mj} of Y_{mj} due to resistances. 95
    - Voltage-controlled current sources 96
  - 6.2.2 Rule to generate the equations 97
  - 6.2.3 Scaling units 104
  - 6.2.4 An example with dependent sources 104
  - 6.2.5 Modifying a circuit 106
  - 6.2.6 Working voltage sources with source transformation . 107
- 6.3 Other considerations for nodal equations 109
  - 6.3.1 Indefinite admittance matrix 109
  - 6.3.2 Superposition in Nodal Analysis 109
  - 6.3.3 Non numerical sources and superposition 110
- 6.4 Programming nodal equations I 111
  - Preparing the inputs 112
  - Programming the calculator 114
- 6.5 Circuits with voltage sources 116
- 6.6 Modified Nodal Analysis (MNA) 118
  - Working MNA with known potentials 122

		6.6.1	Programming Modified Nodal Analysis	123
			Circuit Description	124
			Program pseudocode for mna(n,r,is,gs,vs,ks,ccs)	126
			Developing the program for mna	128
	6.7	"Reducing" number of equations		130
		6.7.1	Using Supernodes	130
			Applying superposition	133
		6.7.2	"Hybrid" reduced MNA	134
		6.7.3	Working with Operational Amplifiers	135
		6.7.4	Closing remarks	138
7	**Loop Analysis**			**139**
	7.1	Loop currents and loops selection		139
	7.2	Resistances and voltage sources only		142
		7.2.1	Circuits with only resistances and independent voltage sources	143
			Scaling units	149
		7.2.2	Current-controlled voltage sources	149
		7.2.3	Indefinite mesh matrix	152
	7.3	Programming mesh equations I		153
		7.3.1	Pseudocode for program	153
		7.3.2	Preparing input for the program	154
		7.3.3	Program for the calculators	155
	7.4	Loop analysis with current sources		157
		7.4.1	Modified Loop Analysis	157
			Remarks on the system of equations	159
		7.4.2	Reducing the number of equations	159
		7.4.3	Hybrid systems	162
	7.5	Superposition and symbolic sources		162
	7.6	Closing remark		162
8	**Network Functions**			**163**
	8.1	Definition of the network functions		163
	8.2	Finding the network functions		165
		8.2.1	Using reduction and transformation	165
		8.2.2	General principles of calculation	167
		8.2.3	Further examples	169
	8.3	Open and short circuit transfer functions		172
	8.4	Closing remarks		174
9	**Superposition: A powerful tool**			**175**
	9.1	Thevenin and Network Equivalents		175
		9.1.1	Basic theory	175
		9.1.2	Finding equivalents: principles	176
		9.1.3	Finding equivalents: examples	177

9.2	Calculating two-port parameters	182
9.3	Finding two-port parameters from equations	182
	Using the nodal and mesh programs	184
	9.3.1 "Active two-ports"	188
9.4	Closing remarks	190

10 Two-port and three-terminal networks   191

10.1	Basic Definitions	191
10.2	Two port parameters	192
	10.2.1 Definition of Parameters	192
	10.2.2 Transforming parameters	194
	Simple cases	194
	Conversion by setting up and solving the equations . .	194
	Programming transformations	196
10.3	Applying two-port parameters	197
	10.3.1 Terminated two-ports	200
	10.3.2 Tandem Connection	203
10.4	Closing remarks	207

11 What's Next   208

11.1	Complex circuits in steady state domain	208
	11.1.1 Elementary notation for complex numbers	209
	11.1.2 Complex numbers and calculator	209
	Settings and Display	210
	Angle setting and polar input	211
	11.1.3 An Example	211
11.2	Short reference to time domain circuits	213
11.3	Closing remarks	215

12 References   216

CHAPTER 1

Preliminary considerations

This chapter offers a brief introduction to the topic of circuit analysis so the organization and procedures in the rest of the book are better understood. The reader may skip the chapter without loss in calculator skills.

1.1 Physical vs. Model Circuits

A *physical circuit* is built with the interconnection of *physical elements or devices*. The size of the physical circuit may be in the nanometers or micrometers range up to hundreds of meters. The number of devices may go from just two, to trillions of elements. In applications of physical electrical circuits we are concerned with the interaction of electrical and magnetic magnitudes such as currents, voltage, electric charge, electric fields, magnetic fields, power, energy and so on.

Unfortunately, it is impossible to fully describe a physical circuit in mathematical terms, or for that matter, even in physical terms. This is true also for simple circuits consisting of just two elements! To deal with physical systems, it is necessary to develop models which are limited descriptions of the devices and their interconnections, focused on some particular characteristics. The validity of an equation and the result or calculation are limited by the accuracy of the models used.

Within an interval of frequencies that goes from DC (0 Hz) up to an upper bound defined by the physical devices in the circuit, an element is modeled by equations referred to as *element equations* which ignore the size and shape of the device. On the other hand, the interconnection obeys to a separate set of equations derived from *Kirchhof's Voltage Law* and *Kirchhoff's Current Law*, which do not depend on devices. The result of this process of modeling, is what we call *model circuits* or simply *circuits*.

How this process fits in real world applications is illustrated in Fig. 1.1. We start from a physical circuit from which, based on the Laws of Physics, physical elements behavior, and other factors to consider, we go through a modeling process and arrive at a model circuit, composed of ideal elements.

In this circuit we apply the methods learned in circuit courses, some of which are reviewed in this book, to arrive at a result. This one has yet to be interpreted so the physical process can be predicted within a certain accuracy. This prediction still needs to be validated by physical experimentation. If validation fails, we first check the analysis methods, and the final interpretation. If the problem is not fixed, then the modeling process and the initial model circuit are reviewed and changed if necessary.

With this figure as a reference, this book, as well as most material in your circuit course, is mainly focused on the methods used to go from the initial model to the result model. In other words, *we work our engineering process with the model circuit, not the physical circuit*. The best proof we have that this methodology is successful is technology working as we know it. In fact, your calculator was designed this way! For this reason, hereafter, the term "circuit" always refers to model circuit.

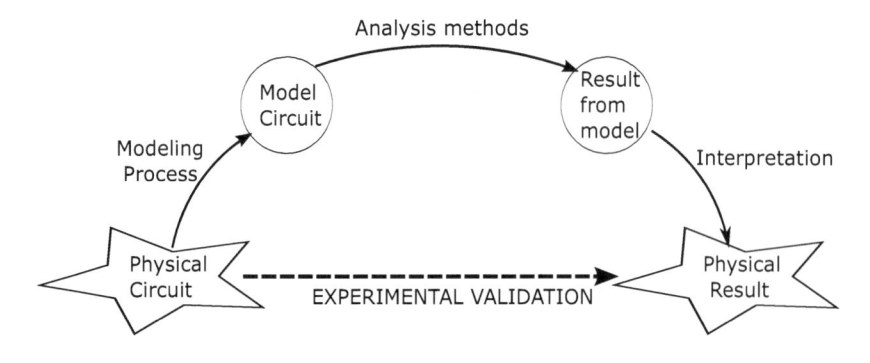

Figure 1.1: Rough description of circuit modeling and analysis.

1.2 Our focus: linear resistive circuits

Let us circumscribe the domain of this book. We deal only with analysis of linear resistive circuits. By analysis it is meant to build up and solve a set of circuit equations by any valid method available. Sometimes our problem does not require all magnitudes. But it is important to understand that no matter the method, when it is necessary we should always be able to find all voltages and currents in the circuit.

To illustrate this remark, the set of nodal equations, for example, solve the node potentials and eventually some other currents. Every other current and voltage not appearing directly in the set of equations must be a function

1.2. OUR FOCUS: LINEAR RESISTIVE CIRCUITS

of these potentials and current variables of our equations. Similarly, when we proceed by circuit reduction, we should be able to go backwards in the process to find any voltages or current.

In circuit courses, we learn how to use the voltage and current in elements to find other magnitudes such as transferred electric charge, absorbed or delivered power and energy, and so on. Sometimes we might introduce an example of these relations, but this is not the focus of the book.

In this book we furthermore limit ourselves to linear resistive circuits with specific elements. Namely, circuits are composed by resistances, linear dependent sources, independent sources, and ideal operational amplifiers. The symbols and equations for these elements are shown in Table 1.1.

Voltage is measured in Volts (V), current in Ampere (A), Resistance R and transresistance r in Ohm (Ω), Conductance G=1/R and transconductance g in Siemens (S) or inverse ohms, $\Omega^{-1} = 1/\,\Omega$.

In sources, the voltage and current at the element are unrelated. The undefined magnitude is determined by its compliance to the respective Kirchhoff's law. In dependent sources, however, the controlling magnitude may be the one in the element, but this is an exception, except for specific applications. In an ideal operational amplifier the output current and voltage are not related to the input magnitudes, but are determined by the equations of connection.

1.2.1 Using this book for AC circuits

Circuits in the phasor domain are not considered in this book. About these circuits, two remarks are worth pointing out.

The first one is of outmost importance, because it makes this book very useful for ac circuts. Namely, the algebraic procedures to find all the phasor voltages and currents in the circuit follow exactly the same methods and procedures that this book covers. An example is shown in the final chapter. With the exception of some properties for resistive circuits, we can affirm that everything presented in the book is applicable to phasor circuits.

The second remark justifies this being the first part of our study. Namely, there are theoretical and practical considerations worth dealing with separately. Also, applications to power systems and circuits in ac domain deserves a treatment by itself.

I invite the reader to take a circuit with complex impedances and phasors for currents and voltages, and apply the methods of reduction, nodal, loop, two ports and others to the circuit. Again, read section 11.1.2 and look at the example there to see what I mean.

1.2.2 Circuit equations and solving methods

At the end, the target in circuit analysis is to solve the circuit equations. The *original* set for a circuit of B elements consists in 2B equations, of which B

Table 1.1: Linear Resistive Elements

Element	Symbol	Equation
Resistance		$V = RI$ or $I = GV$
Independent Voltage source		V given
Independent Current source		I given
Voltage Controlled Voltage source (VCVS)		$V = kV_x$
Voltage Controlled Current source (VCCS)		$I = gV_x$
Current controlled Voltage source (CCVS)		$V = rI_x$
Current controlled Current source (CCCS)		$I = \alpha I_x$
Ideal Operational Amplifier (OA)		$V = 0$ $I_1 = I_2 = 0$

1.2. OUR FOCUS: LINEAR RESISTIVE CIRCUITS

are derived from Kirchhoff's laws, and the other B equations correspond to elements. To tell you the truth, these are too many equations!

To simplify this task, different methods have been developed. These can be divided in two great groups: a) working by reduction and transformation, and b) setting up a reduced number of equations following some rules.

In reduction and transformation methods, sub circuits are identified and substituted by simpler or smaller equivalent configurations, perhaps one element, in such a way that the new circuit is smaller or easier to deal with. We could say that these methods are akin to reducing the number of equations and variables. The procedures in this group depend on your ability at identifying structures and applying your intuition. They help you to develop, on the other hand, your intuition for easier recognition of properties and structures, and for design purposes. Some structures are so common that they are worth programming.

The other group consists in using systematic methods of establishing sets of equations, which are in a smaller number than the 2B original equations. The most popular among these methods are the nodal and loop analysis methods. These are the ones usually taught in undergraduate courses and the ones we deal with in this group. There are other methods that the reader may consult elsewhere.

The methods in the second group are more suitable for mathematical algorithms, because they follow well established step by step processes. As a consequence, they are easy to program.

1.2.3 Hand analysis vs analysis with calculator

Many students I know use the calculator as an extension of their pencil. That is, just to do operations. Before following this path, it is worth looking at what your calculator has and how working with it will be different.

When working with pencil and paper, you try to use simpler formulas and simplify calculations before proceeding further. This method is not necessarily the best when using calculators. For example, when combining three or more resistances in parallel by hand, it's easier to work in steps of two resistances per round. Why? Because multiplying R1 and R2, and then dividing by (R1+R2) is more intuitive than adding the inverses $1/a$ and $1/b$, and then taking the inverse again. In fact, I've found "programs" in internet that work in this way for three or more resistances.

But in the calculator it is easier to work $(1/a + 1/b + 1/c)^{-1}$ or $1/(1/a + 1/b + 1/c)$ directly! One formula for any number of resistances in parallel.

This is just one of many examples that tell us that you must do your homework in setting your mindset for calculator use. Know your calculator and your theory, and look for methods that let you apply your calculator effectively. This principle applies to any technological tool, including software like Matlab^{TM}or Mathematica^{TM}, and so on.

1.2.4 Programming and using programs

I repeat what I already said before: " if you don't know how to do it with pencil and paper, you don't know how to do it with the calculator". Running a program doesn't mean you work by yourself the task the program is doing. Even assuming it is well written.

To illustrate my point, if you enter the data of the circuit to the SPICE simulator, you will get the node potentials. You did not calculate the node potentials, and SPICE did not learn you how to do it. If you don't know how to obtain those potentials, it is not by using SPICE that you will learn the method.

My first advice here is then: do not attempt to write down a program in your calculator until you are sure you have learned yourself how to solve the problem. Of course, once you know how to do things, write the program and use it! You have earned that privilege! Or you may download one from the internet for that matter, or from your friend's collection. On the other hand, planning and writing a program may be an adventure by itself, because at the end you will follow paths suitable for the calculator that would not be used with pencil and paper. It is worth the experience. And the prize is very rewarding.

CHAPTER 2

Practical matters.

In this chapter I explain several points related to the notation used to simplify writing for me, and reading for you, as well as remarks relating to features of the calculator that will allow us to make better use of them. You do not need to read this chapter first if you have already a good understanding of your calculator model. Also, you may refer to the chapter whenever you need something in particular.

2.1 Texas Instruments calculators used in this book

This book does not intend to be a handbook for the the TI-nspire cx , illustrated in Fig. 2.1. The reader may consult what is in my opinion a good introduction to it at

https://education.ti.com/sites/UK/downloads/pdf/First_Steps_with_TI-Nspire_CX.pdf

for more information. For our purposes, the TI-Nspire CX family includes the TI-Nspire with touchpad, TI-Nspire CAS, TI-Nspire CX, and TI-Nspire CX CAS.

Texas Instruments has a practical emulator which you may install and use for anything you want. But I will follow here with the principle that you are using your calculator because you carry it with you. And sometimes you want to do something and do not have your emulator available. Let us now look at some features that we may need.

2.2 Keys and key typing

Let me introduced now some remarks on notation for keys.

8   CHAPTER 2. PRACTICAL MATTERS.

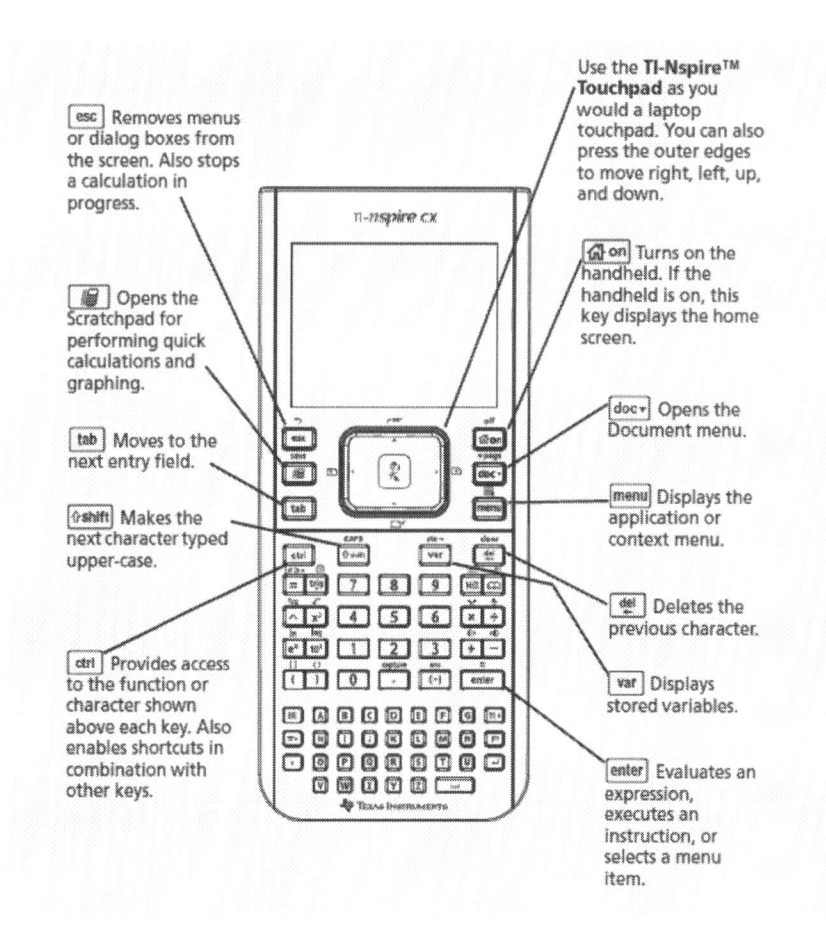

Figure 2.1: TI-nspire cx Key layout (*Image Courtesy of Texas Instruments, Inc.*)

2.2.1 Cursor keys

The cursor in the graphing calculators may be moved along up, down, left or right directions using the cursor keys. These are small triangles located on the touchpad: ◄, ►, ▲ and ▼. The cursor itself appears blinking. If you hold the shift key down while moving a cursor key, you will highlight a word or entry, and you will be able to copy it (ctrl C) or cut it (ctrl X), so you can paste it (ctrl V) later.

I use the triangles with no frame to indicate pressing a cursor key.

2.2.2 Key notation and key values

As for keys, sometimes or when necessary I write the face character within a frame. This notation tells you the key to press, not the value you get. For example, with $\boxed{\texttt{=}}$ you get the equal symbol "=", but the sequence $\boxed{\texttt{ctrl}}$ $\boxed{\texttt{=}}$ opens a menu of relational symbols to select from. Similarly, $\boxed{\pi \blacktriangleright}$ opens another menu. I describe these menus later.

Since letters A to Z, and digits 0 to 9 do not have secondary values, I will not frame them. I usually do not use the frame for the operation symbols $\times$, $\div$, $+$, $-$, and $\wedge$. Although $\boxed{\texttt{EE}}$ does not have a secondary value either, I use the frame to avoid confusion. Similarly with $\boxed{\texttt{enter}}$ and $\boxed{\texttt{(-)}}$. In all cases I use `typewriter font` to indicate the sequence to be entered.

2.2.3 Remarks about menu keys

Menu keys are those that, either as a face value or a secondary value, they actually open a menu to select from. You navigate the menu with the cursor keys or else by pressing again the same key (not always). Once your item is highlighted, you type $\boxed{\texttt{enter}}$ to take it to the command line. To exit the menu without making a selection, hit $\boxed{\texttt{esc}}$.

The menu keys are five: $\boxed{\texttt{trig}\blacktriangleright}$, $\boxed{\pi \blacktriangleright}$, $\boxed{\texttt{?!}\blacktriangleright}$, the templates key $\boxed{\vcenter{}}$, and the catalog key $\boxed{\varpi}$ with an open book[1].

The sequences $\boxed{\texttt{ctrl}}$ $\boxed{\texttt{=}}$ and $\boxed{\texttt{ctrl}}$ $\boxed{\varpi}$ also open menus. I suggest the reader to open the menus and recognize the different symbols and items.

2.2.4 Grouping symbols and entry fields

When calling functions such as `sin()`, `ln()` and so on, the cursor will stay within the parenthesis until you exit from it. Grouping symbols open by pairs and function similarly. These symbols are the brackets [ ], with $\boxed{\texttt{ctrl}}$ $\boxed{\texttt{(}}$; braces { } with $\boxed{\texttt{ctrl}}$ $\boxed{\texttt{)}}$, and parentheses. To exit those fields, you press the right arrow key on the touchpad or hit the $\boxed{\texttt{tab}}$ key. For example, to produce and execute the expression {3*π,8} + 5 you enter the sequence

$\boxed{\texttt{ctrl}}$ $\boxed{\texttt{)}}$ 3 $\boxed{\pi \blacktriangleright}$ $\boxed{\texttt{enter}}$, 8 $\blacktriangleright$ + 5 $\boxed{\texttt{enter}}$

or $\boxed{\texttt{ctrl}}$ $\boxed{\texttt{)}}$ 3 $\boxed{\pi \blacktriangleright}$ $\boxed{\texttt{enter}}$, 8 $\boxed{\texttt{tab}}$ + 5 $\boxed{\texttt{enter}}$

Same remarks apply to the fields for the exponential function keys $\boxed{e^x}$, $\boxed{10^x}$, $\boxed{\wedge}$, the square root with $\boxed{\texttt{ctrl}}$ $\boxed{x^2}$ and so on.

As a final remark, division can be entered as it is done traditionally or using the fraction secondary value $\frac{\square}{\square}$ obtained with the sequence $\boxed{\texttt{ctrl}}$ $\boxed{\div}$.

---

[1] I really tried to make it look like an open book. ;)

The two sequences below produce the same result on the screen, $\frac{3+2}{8-6}$:

$\boxed{(}\ 3\ +\ 2\ \blacktriangleright\boxed{\div}\ \boxed{(}\ 8\ -\ 6\ \blacktriangleright\boxed{\text{enter}}$
$\boxed{\text{ctrl}}\ \boxed{\div}\ 3\ +\ 2\ \blacktriangledown\ 8\ -\ 6\ \blacktriangleright\boxed{\text{enter}}$

Do it as you prefer.

2.2.5 Powers of 10 using EE

When you want to enter a datum that includes a power of 10, such as 1.2345×10^4, 3567.5×10^{-6}, you may use the $\boxed{\text{EE}}$ key instead of the ten's power $\boxed{10^x}$. For example, these data may be entered as follows:

1 . 2 3 4 5 $\boxed{\text{EE}}$ 4   and   3 5 6 7 . 5 $\boxed{\text{EE}}$ $\boxed{(-)}$ 6

On the window, the exponent appears after the letter E: `1.2345E4`

2.2.6 Notation for entry and its result

Let me introduce the notation used in the book regarding the inputs on the command line, and what results from the input.

To show the result on the stack from an entry line after pressing $\boxed{\text{enter}}$, I use a right arrow, $\to$. Thus, the next expression means that 3+5 on the command entry line yields 8 after pressing the key $\boxed{\text{enter}}$.

3 + 5 $\boxed{\text{enter}}$ $\to$ 8

Sometimes the key $\boxed{\text{enter}}$ is omitted if no ambiguity arises.

3 + 5 $\to$ 8

On the other hand, when a sequence of keys are required to call a function or a program, I use a $\Rightarrow$ to show how the procedure produces the name called.

By the way, predefined or user defined functions or programs may be also be invoked by directly writing the name on the command line line.

2.3 Scratchpad and documents

The TI-nspire gives you the option to work on a document or to work on the Scratchpad. This one is used, to say it somehow, for calculations that are done on the spot or that you can erase without remorse. You can, however, define functions here as explained later in section 2.7.

You cannot, however, program while in the scratchpad. For that purpose, you need a document. I suggest you to create one where you can gather all your programs and functions. For example, let us create the document

2.4. SETTINGS FOR DISPLAY AND DATA INPUT 11

"CIRCUITS" to store our programs. Turn on your calculator and hit first the following sequences:

1 1 1 5

With these steps you have created a document with program editor included, and opened the dialog window shown in Fig. 2.2

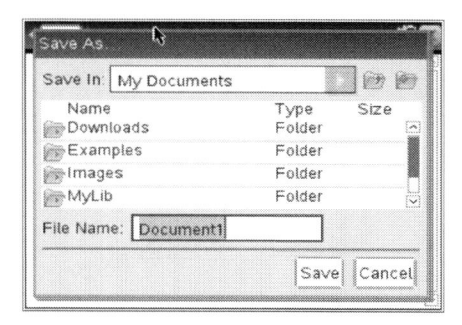

Figure 2.2: Dialog window for saving documents (*Image Courtesy of Texas Instruments, Inc.*)

Type the name of the document and use the tab key to change fields. Once on "Save", press enter. Now you can continue working here as in any calculator or you may open the Scratchpad hitting the key with a flying calculator, 🖩. Notice that any variable that you define on the document is local to it. Ditto for the Scratchpad.

2.4 Settings for display and data input

Now, let us look at the mode settings in the calculator which determine how data results are displayed and, in some cases, also read from the entry line. For our purposes, we concentrate in the settings for number display, angle units, and for complex data display. Most settings, not all, affect only the display of the answer, not the input format of data.

To define your settings, open the Documents Settings window with the sequence

doc▼ 7 2

Once all settings are set, a final enter takes you back to the environment you called the settings from.

On the Documents Settings window, use the up and down cursor keys to chose the settings menu to change. The right navigation ▶ takes you then to the menu options from where you select the desired one. If you want to go back without changing the item, you may hit [esc] or the left navigation key ◀. Once your item is highlighted, press [enter] and proceed to the next setting.

2.4.1 Numeric Display Formats:

The settings used to establish how results will be displayed are: a) Display Digits, b)Exponential Format, and c) Calculation Mode. In addition we are interested in the angle and the Real or Complex settings.

"Display Digits" menu

Is used to control the number of digits to be displayed, or else the number of digits in the decimal part.

The option *FLOAT* alone shows the total number of digits displayed and the digits in the decimal part varies, depending on the result. Now, for the other options:

Float N: *FLOAT N*, where integer $N \geq 1$, will display results rounded to a maximum total number of N digits. If necessary, will automatically switch to scientific mode to maintain the N digits.

Fix N The option *FIX N*, $N \geq 0$, will determine the number of digits in the decimal part. Notice that if you select FIX format, then extra 0's may be added to the decimal part if necessary.

Exponential Format menu

The exponential format deserves particular attention because when used wisely our results are easier to read and faster to interpret. In all options the number of digits are displayed according to the settings described before.

Normal: result shown in full. When not exact, the least significant digit is rounded.

Scientific: Result is displayed in the form n.ddddE*M*, where $0 < |n| < 10$. It represents the normal scientific notation $n.ddd\ldots \times 10^M$. For example, 2.4567E-2 corresponds to 2.4567×10^{-2}. Normal displays may switch to the scientific format automatically when necessary.

Engineering: Similar to the scientific format, except that $0 < |n| < 1000$ and M is always a multiple of 3.

The above descriptions stand for the displayed results, not for the input. You may enter the numbers in any format you want. If we are working in scientific mode display, datum 2.4567×10^{-2} may be entered as 2.4567 [EE]

2.4. SETTINGS FOR DISPLAY AND DATA INPUT

`(-)` `2`, `0.024567`, `245.67` `EE` `(-)` `4`, or any other form. In all cases the displayed result will be the same.

One advantage of the engineering notation is that the powers of ten are directly related to the prefixes used in engineering and shown in Table 2.1.

Table 2.1: Prefixes for the SI units

Prefix	symbol	Power of 10	Display
femto	f	10^{-15}	E-15
pico	p	10^{-12}	E-12
nano	n	10^{-9}	E-9
micro	μ	10^{-6}	E-6
mili	m	10^{-3}	E-3
N/A		10^{0}	E0
kilo	k	10^{3}	E3
mega	M	10^{6}	E6
giga	G	10^{9}	E9
tera	T	10^{12}	E12

Prefixes may be adapted to situations where units are scaled. For example, talking of current in Amperes, 2E-3 means 2 mA. Yet, if the current in calculation is already in μA units, then 2E-3 will mean 2 nA.

This feature makes it easier to read and interpret results. We will have opportunities to illustrate this remark.

Calculation menu

The menu has three options: *Auto, Exact, Approximate*, described next.

Exact Any result that is not a whole number is displayed in a fractional or symbolic form ($1/2$, $3\sqrt{2}$ etc.). However, if the input contains a decimal point, the result is displayed in approximate mode. Hence, 3/6 produces 1/2, but 3/6. yields 0.5.

You can translate an exact expression to an approximate one with the sequence `ctrl` `enter`.

Approximate All numeric results are displayed in floating-point(decimal) form, i. e., an integer or a number containing a decimal point.

Auto Provides the EXACT form where possible, but the APPROXIMATE form if the entry contains a decimal point.

I usually prefer the *APPROXIMATE* feature, and this will be default in this book. Whenever the result is not exact, the least significant digit is always rounded.

2.4.2 Angle mode

This mode sets the units in which angle values are interpreted when input and displayed. Three options are offered: *RADIAN (Rad), DEGREE (Deg) or GRADIAN*. If you work with trigonometric functions or complex numbers, this is an important setting to define. For inputs, the setting may be overridden with the ° feature if you want to enter angles in degrees, ^{r} if the angle is in radians, and ^{g} if it is in gradians. These symbols appear in the $\boxed{\pi \blacktriangleright}$ menu key. For example,

$\sin(30) = 0.5$ in degree mode, $\sin(30) = -0.988$ in radian mode, and $\sin(30) = 0.454$ in gradian mode, but $\sin(30°) = 0.5$ in any mode.

2.4.3 Real or Complex menu

In the *real* setting, results should always be real, unless a complex number is enter. Thus, in real mode $2\sqrt{-1}$ yields an error message, but $2i$, where i is the imaginary unit equal to $\sqrt{-1}$ yields $2i$.

When working with complex numbers, you must select rectangular or polar settings.

When in *Rectangular* mode, complex results are displayed in rectangular format $a + bi$

When in *Polar* mode, complex results are displayed in polar format $a\angle\phi$ when angle mode is in degrees, and in the polar format $e^{i\phi} \cdot a$ when in radian mode.

Regarding the input of an angle ϕ, the calculator will use the angle settings being used.

The input of complex numbers may be in any format. An input in polar form using the $\angle$ symbol must be enclosed in parenthesis, $(A\angle\phi)$. When the angle settings is in radians, you have the option of using the exponential form for the input, $Ae^{i\phi}$. You can mix numbers in different formats. For example, assuming radian mode for the angle and polar format for display:

$$\frac{(3+2i) * (4 \angle .38)}{6 * e^{0.56i}} + 1.4 - 0.79i \rightarrow e^{0.0454 \cdot i} \cdot 3.6101$$

If you try the same input in degree mode, the calculator will show an error because of the invalid exponential polar input used.

2.5 Variables and Storage

Variables are important for our goals, so let me stop a little to discuss about them in the TI-nspire cx.

Variable names The name may have any number of letters or digits, but must start with a letter

2.5. VARIABLES AND STORAGE

Unassigned variable If a variable does not have a "value" assigned, it is treated as an algebraic symbol.

- Example: if no value has been assigned to X, then
  x → x
  x∧2 + 9 → x^2 + 9

Assigning value The TI-nspire cx has three methods to assign a value to a variable.

1. With the "store" function. This is a secondary value of key $\boxed{\text{VAR}}$. To assign value 4 to X:
   4 $\boxed{\text{ctrl}}$ $\boxed{\text{var}}$ X. To simplify, I write 4 $\boxed{\text{sto→}}$ X

2. With the "Define" function. You may enter directly the name using the keypad, or call it with the sequence $\boxed{\text{menu}}$ 1 1. To assign value 4 to X
   $\boxed{\text{menu}}$ 1 1 X = 4. I write simply Define X=4

3. With the "assign operator" := which you find as secondary value of key $\boxed{}$. To assign value 4 to X:
   X $\boxed{\text{ctrl}}$ $\boxed{}$ 4. To simplify I write X := 4

Valid values The variable can be assigned different types of values: numbers, lists or matrices (see next chapter), algebraic expressions with unassigned variables (see subsection 2.7.1), strings, and others.

2.5.1 The $\boxed{\text{var}}$ key

This key displays a list which contains the variables you have defined. Use the up and down cursor keys to highlight a specific variable. When you press $\boxed{\text{enter}}$, this variable will be pasted onto the command line.

The list shows not only the variable name, but also has icons to let you know what type of variable it is.

Deleting variables: Using the Actions menu you can delete variables, except function variables. To call this menu press $\boxed{\text{menu}}$ 1, and select your action. Then,

1. **With 4: Clear a-z** ($\boxed{\text{menu}}$ 1 4) you delete all variables named with only one letter, including function variables.

2. **With 3: Delete Variable (DelVar)** ($\boxed{\text{menu}}$ 1 3) After calling the action, select your variable using $\boxed{\text{var}}$.

3. **To delete a function variable:** Press $\boxed{\text{menu}}$ 1 3, select your function variable, and hit $\boxed{\text{del}}$ to delete parenthesis. Then $\boxed{\text{enter}}$

2.5.2 Browsing the stack

You can recall entries or results using the up and down cursor keys to navigate through the stack, pressing [enter] once you highlight the desired item. If you select a previous entry, this one is copied onto the entry line, where you can edit it. If you select a previous result, this value will be pasted onto the entry line.

2.5.3 Variable ANS

A secondary value of the key [(-)] is the variable ANS. This variable stores the last result and is updated after each transaction. One of many advantages of using ANS is in reducing rounding errors.

This variable has an important characteristic. If you start the command line directly with an operation, the assumed operand is ANS. The following lines illustrate this comment. Each line shows first what you enter on the entry line, and to the right what it actually appears on the line. Some lines would not apply to your calculator because of the lack of the key:

+ 2 $\Rightarrow$ ans + 2

[∧] 2 $\Rightarrow$ ans^2

[÷] 2 $\Rightarrow$ ans / 2

There are many uses of this variable. One of them is to help us in breaking a complex formula for calculation. Let us illustrate this with an example.

Example 2.1 *The objective is to calculate the expression*

$$M = \frac{1.26 \times 10^3 (287 + 33.25 \times 10^2)}{1.26 \times 10^3 + 287 + 33.25 \times 10^2}$$

The expression in parenthesis on the numerator is calculated first. Settings in Engineering mode. The sequence is as follows:

287 + 33.25 [EE] 2 [enter] $\rightarrow$ 3.612E3

[×] 1.26 [EE] 3 [÷] (1.26 + ANS(1)) $\rightarrow$ 934.138E0

The variable can also be used to do iterations, as shown in the following example. We use here the fact that pressing [enter] directly, the last command line is repeated.

Example 2.2 *We want to approximate one real root of $f(x) = x - \cos(x)$ using Newton's method to ten decimal places. The iteration principle for the method is the formula*

$$x_{n+1} = x_n - \frac{f(x_n)}{f'(x_n)} = x_n - \frac{x - \cos(x)}{1 + \sin(x)}$$

We start start setting FIX 10, Normal *modes to have the desired number of decimals. Let us start our guessing with x=-1, and introduce Newton's formula in the entry, assuming* ANS(1) *is the current value* x_n:

`(-)` `1` `enter` → -1

`ans-(ans-cos(ans)` `÷` `(1+sin(ans))` `enter` → 8.7162169588

`enter` → 2.9760606551

`enter` → -.4257846887

`enter` → 1.8511838176

`enter` → .7660395196

`enter` → .7392410675

`enter` → .7390851386

`enter` → .7390851332

`enter` → .7390851332

Since the last value repeated itself, the sequence has converged within ten places to the approximate solution 0.7390851332.

2.6 Using the colon (:) in entries

In just one entry line you can enter several commands, separating them with the colon (:). This character appears in the `?!▶` list, and can be retrieved with the sequence `?!▶` `?!▶` `▼` `enter`. Only the last result will appear on screen and be saved in the ANS variable.

This feature is convenient for definitions, to reduce lines on screen, and many other situations. I provide an example here where this feature is combined with the property of the `enter` key which repeats the last entry if hit, to solve a set of non linear equations by iteration[2]

---

[2]I have chosen an example that converges. Theoretical discussions on convergence are outside our scope

Example 2.3 *Let us solve the following system to four decimal places by iteration. Set the system to angle in Radians and display to FIX 4.*

$$x + 3\sin(y) = 1.6$$
$$2\cos(x) + 1.5y = 3.4$$

Start with the value y=0. In the entry below, I use the list {x, y} to display the values.

```
y:=0
x:=1.6 - sin(y):y:=(3.4-2cos(x))/1.5:{x,y}  enter  → {1.6000,2.3056}
 enter  → {0.8580,1.3928}
 enter  → {0.6155,1.1780}
 enter  → {0.6762,1.2267}
 enter  → {0.6586,1.2122}
 enter  → {0.6636,1.2163}
 enter  → {0.6622,1.2151}
 enter  → {0.6626,1.2155}
 enter  → {0.6625,1.2154}
 enter  → {0.6625,1.2154}
```

Since the last set repeted, we have to four decimal places x = 0.6625 *and* y=1.2154. *The reader can verify these results by substituting in the original equations.*

2.7 User Defined Functions I

In a way, functions, both those included in the calculator and those defined by the user, behave like variables that can be used in your computations and programs as any other variable defined previously. Except that the function may depend on other variables.

In the TI-nspire cx, functions may be defined either on the command entry line (I prefer this method for simple ones) or with the program editor. In the first case, you can define them in the Scratchpad or in your document. In the second case, you must do it in your document. In any case, the name is included in the variable list called with var . If your function is too complex, it may be better to use the program editor. I give examples later.

There are two classes of functions that you can define on the command entry line: *algebraic non parametric* functions, and *algebraic parametric* functions. In the first group, the variable is not protected. For the second group, the definition can be edited, but the variable cannot be assigned a new value without previously deleting it.

2.7. USER DEFINED FUNCTIONS I

2.7.1 Algebraic non-parameter functions

Since any undefined variable behaves like an algebraic entity, an expression involving one or more unassigned variables (x1, x2, ...) stored onto another variable y, [expression sto→ y], makes y a function of those variables. When you assign numerical values, y is modified accordingly.

The variable y is not protected and may be assigned other expression or value. See the following examples.

Example 2.4 *With variable x not assigned, define the "function" $y = x^2 + 1$ as shown in the first line of the table below. In the second line, x is assigned the value of 2, and in line 3 we check that y's value is updated. Lines 4 and 5 repeat the task for x=4. Next, the value 8 is directly assigned to the unprotected variable y, and the functional relationship is lost, as illustrated by the last two lines.*

Line	Entry	Answer on stack
1.	x∧2+1 sto→ y enter	$x^2 + 1$
2.	2 sto→ x enter	2.
3.	y enter	5.
4.	4 sto→ x enter	4.
5.	y enter	17.
6.	8 sto→ y enter	8.
7.	2 sto→ x enter	2.
8.	y enter	8.

Notice that initially you define a one letter function y of x. Each time you assign a value to x, that of y is updated. Once you assign a value to y directly, the functional relation is lost. The same remarks apply for a function of several variables, as illustrated next:

Line	Entry	Answer on stack
1.	a2+ b1-2 sto→ y enter	$a1^2 + b1 - 2$
2.	4 sto→ a1 enter	4.
3.	y enter	b1 + 4.
4.	2 sto→ a1: 5 sto→ b1 enter	5.
5.	y enter	17.
6.	8 sto→ y enter	8.
7.	y enter	8.

Notice that if you assign a numeric value to one of the variables (a1 in line 2) while the other is still unassigned, update of y is done only with the assigned variable (line 3).

This feature is very handy for those situations in which you have to calculate a formula for different values which are not known beforehand, including lists or matrices,[3] but there is no need for a permanent function. For a permanent function, go to the following subsection.

As a matter of fact, notice that on line 4 of the second sequence we use the colon (:) to write two entries on the command line. Only the last entry goes to the stack, while other commands are executed but not shown or stored in stack.

2.7.2 Functions with parameter

Defining explicitly the function with parameter, y(x) or y(a1,b1), makes the parameters local variables. When calling the function, you substitute the desired values inside the parenthesis. However, now variable y is no longer available for other uses. If you want to free this variable, you must delete it (see section 2.5.1). On the other side, the parameter variables x, a1, b1 in the calculator are independent and assignments do not affect y.

A function variable can be edited by recalling it with the sequence menu 1 2 enter, and then selecting the variable.

Example 2.5 *The function $y(x) = x^2 + 1$ is defined with the parameter x, and the previous remarks illustrated by the following lines. The function definition uses an unassigned variable x.*

Line	Entry	Answer	Comment
1	y := x^2+1 enter	done	Define function
2	y enter	error message	Variable y does not exist by itself
3	x:= 6 : y(4) enter	17.	x=4 locally
4	8 sto→ y enter	error message	y is protected

2.8 Programming

You can create either programs or functions with the program editor. The main difference is that the function returns a value that can be used as any other variable in operations, while the program must be executed independently. Thus, for the function pl(z), a command such as 35 + pl(z) makes sense and is executed. But if it is a program, it will not work.

To illustrate the process and execution, let us create two programs. The theory behind them is found in the formulas (5.11) and (5.12) on page 65 for the wye-delta and delta-wye transformations. The steps to create, save and test your program are as follows:

---

[3]If you know the values beforehand, you don't need a function. A list operation will do!

2.8. PROGRAMMING

1. With your document open (I work in "Circuitos", but you use your own) type [menu] 9 1 1. This will open a dialog window to create your program. Enter the name "delta2wye". At this point you have something as shown in Fig. 2.3(a). With the [tab] key you change fields if you need so.

2. Hit [enter] and your page will be partitioned in two sub windows, as illustrated by Fig. 2.3(b). One is the program editor. You may switch between these sub windows with [ctrl] [tab].

3. In the program editor field type your program. (See example below) Then press [menu] 2 1 to save and store your program. Save your document with [ctrl] S. If you hit [var], you should see the program in the list of variables.

4. At this point, it may be a good idea to test your program, before closing the editor. Switch fields with [ctrl] [tab], and run your program. You may type the name or call it with the [var] key.

5. If you are satisfied, you may close your editor. To do so, switch fields first. Then press [menu] 1 D to close it.

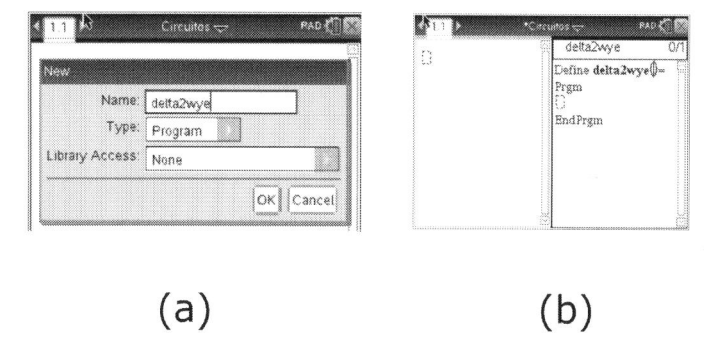

(a) (b)

Figure 2.3: Dialog windows to create a program (*Image Courtesy of Texas Instruments, Inc.*)

Let us now go to an example.

Example 2.6 *The first program shown in Fig. 2.4 on the next page. In this figure, only the parts in* typewriter font *constitute the program. I do not include comments for brevity. Line numbers are for reference and comments are shown in the right column for explanation.*

This program gets the Y or T equivalent values for a delta subcircuit (see Section 5.2). The formulas used are

$$R_1 = \frac{R_{12} R_{13}}{R_{12} + R_{13} + R_{23}} \qquad (2.1a)$$

$$R_2 = \frac{R_{12}\, R_{23}}{R_{12} + R_{13} + R_{23}} \tag{2.1b}$$

$$R_3 = \frac{R_{13}\, R_{23}}{R_{12} + R_{13} + R_{23}} \tag{2.1c}$$

After typing the program and saving it with [menu] 2 1, let us test it. We switch fields hitting [control] [tab] and either type *delta2wye()* or else retrieve it from the variable list. After testing the program you may switch fields again and close the editor with [menu] 1 D.

```
1    delta2wye()                         Comments
2    Pgrm
3    Local s, r12, r13, r23
4    Request ''Enter r12:  '', r12      Requesting input
5    Request ''Enter r13:  '', r13
6    Request ''Enter r23:  '', r23
7    r13 + r12 + r23 → s                sum in denominator
8    r12*r13/s→ r1                      Finding and storing
9    r12*r23/s→ r2                      wye's resistances
10   r13*r23/s→ r3
11   Disp ''r1 ='', r1                  Displaying results.
12   Disp ''r2 ='', r2
13   Disp ''r3 ='', r3
14   EndPrgm
```

Figure 2.4: Delta-Wye Transformation Program without passing parameters

As a second program, let us write the reverse conversion, from wye to delta with a program that will now receive the input as parameters. The formulas are

$$R_{12} = \frac{R_1\, R_2 + R_1\, R_3 + R_2\, R_3}{R_3} \tag{2.2a}$$

$$R_{13} = \frac{R_1\, R_2 + R_1\, R_3 + R_2\, R_3}{R_2} \tag{2.2b}$$

$$R_{23} = \frac{R_1\, R_2 + R_1\, R_3 + R_2\, R_3}{R_1} \tag{2.2c}$$

The program is shown in Fig. 2.5. To run it, you may call it from the variables list and introduce your input parameters in the parenthesis. For example wye2delta(130e3,125e3,28e3).

2.9. CLOSING REMARKS

```
1    :wye2delta(r1,r2,r3)           Comments
2    :Pgrm
3    :Local s
4    :r1*r3 + r1*r2 + r2*r3 → s     sum in numerator
5    :s/r1→ r23                     Finding Delta's resistances
6    :s/r2→ r13
7    :s/r3→ r12
8    :Disp ''r12 ='', r12           Displaying results.
9    :Disp ''r23 ='', r23
10   :Disp ''r13 ='', r13
11   :EndPrgm
```

Figure 2.5: Wye-Delta Transformation Program passing parameters

In most programs, unless there is some specific problem, you can enter or call the program using assigned or unassigned variables. As an example, I suggest you to run `delta2wye()` and, when windows open enter as data unassigned variables, for example `a, b, c`. You cannot use however the same unassigned variable notation `r12` because an error is detected.

Furthermore, you may enter lists as data or parameters. Lists are a topic of next chapter.

Also, in the above, and similar programs or functions, when you use dot operations, matrices of equal dimensions are acceptable as data. The reader may try it.

Examples for defining functions using the program editor will be given in later chapters.

2.9 Closing remarks

This chapter does not pertain to circuit analysis (except for examples), of course. But it provides us with information about your calculator which will be useful for the rest of the book. Many students have found this chapter very useful.

Many readers will be able to find interesting and useful features that I have not included, either because I did not considered them worth talking about them – wrongly, perhaps – or else simply because I did not know them. If you know of one feature that should be included, don't hesitate to tell me. I thank you in advance.

CHAPTER 3

Lists and Matrices

This chapter provides a quick review of some basics for two tools that will be very important for us: lists and matrices. As the reader might already be thinking, for our case matrices are important tools for solving sets of linear equations.

3.1 Lists

Readers who have used the calculator for Statistics may be already familiar with lists, also called arrays. In fact, the association is so strong that in the TI-nspire cx the list operations are all in the Statistics menu! But lists are indeed excellent tools for our goals, and therefore I will devote a special section to them.

A list of *dimension* n is an ordered array of n numbers or algebraic expressions between braces ({ }), separated by commas, like

$$L = \{x_1, x_2, \ldots, x_n\} \qquad (3.1)$$

A list can be defined or edited using the command line or an editor. To take full advantage of lists *always* store them as variables. Individual elements in the list may be entered with operations, leaving the required computation to the calculator. However, if the individual calculations are complicated, it may be convenient to store them first in memory.

Examples of lists, their definition are shown next. Remember, for $\{2,3,4\}$ `sto→` L1, you actually type `ctrl`) 2,3,4 ▶ `ctrl` `var` L 1

3.1. LISTS

L1 = {2, 3, 6}
  ENTRY: { 2 , 3 ,6 } [sto→] L1
L2={5.18, 3.98, 6.59, -0.245}
  ENTRY: { 5.18 , 3.98 , 6.59 , [(-)] 0.245} [sto→] L2
L3 = {5+3.2, 6/1.76, 3^2}
  ENTRY: { 5 + 3.2 ,6 ÷1.76,3 [x^2] } [sto→] L3

3.1.1 Working with Lists

Now that you know how to create a list and store it onto a variable, play around with the operations.

A) Operations with a scalar: If a is a scalar, i. e., a number, and * denotes an operation with numbers, then for $L = \{x_1, x_2, \ldots, x_n\}$, we have

$$a * L1 = \{a * x_1, a * x_2, \ldots, a * x_n\} \quad (3.2)$$
$$L1 * a = \{x_1 * a, x_2 * a, \ldots, x_n * a\} \quad (3.3)$$

Example 3.1 *Look at the following expressions*
```
3+{-3, 2, 5}    →    {0.     5.      8};
3÷{-3, 2, 5}    →    {-1.    1.5.    0.6};
{-3, 2, 5}^3    →    {-27.   8.      125};
{-3, 2, 5}/3    →    {-1.    0.667.  1.667}
```

B) Operations with lists: Let us take two lists of equal dimension, $L1 = \{x_1, x_2, \ldots, x_n\}$ and $L2 = \{y_1, y_2, \ldots, y_n\}$, and the asterisc (*) represent a binary operation for numbers. Then

$$L1 * L2 = \{x_1 * y_1, x_2 * y_2, \ldots, x_n * y_n\} \quad (3.4)$$

whenever the individual operations are allowed.

Example 3.2 *Take* L1 = {2, -3, 4}, L2={-1. -2. 5.}. *Then,*

```
L1+L2    →    { 1.     -5.      9.}
L1^L2    →    = {0.5   0.11111  1024.}
L1/L2    →    {-2.    1.5       0.8}
```

C) Functions: If $f(x)$ is a function and $L1 = \{x_1, x_2, \ldots, x_n\}$ is a list, then

$$f(L1) = \{f(x1), f(x2), \ldots, f(xn)\} \quad (3.5)$$

That is, the function is applied to each element of the list. Look at the following examples.

Example 3.3 *(1) Being in radian mode, $L = \{0, \pi/4, \pi/2, -\pi/4\}$ then*

$$\sin(L) = \{0, 0.7071, 1, -0.7071\}$$

(2) Working in exact mode, let

$$f(x) = \frac{\sqrt{x}}{2+x} \quad \text{and} \quad L = \{4, 9, -16\}$$

Then

$$f(L) = \{1/3, 3/11, -2i/7\}$$

As you might have noticed, user defined functions work fine with lists.

Example 3.4 *For a voltage divider with two resistances R1 and R2, and a source Vs, the formula for the output voltage is given by*

$$V_o = \frac{V_s\, R_2}{R_1 + R_2}$$

Let us assume that we want to find the outputs for the different combinations of values in the next table.

Vs (V)	R1 (Ω)	R2 (Ω)	Output (V)
5	600	2100	
8	1220	3900	
12.7	655	590	
3.1	320	780	
6	1560	4820	

Let us first define the function:

Define vo(vs,r1,r2) = vs · r2 / (r1 + r2) [enter] → Done

We can now define the lists:

{5,8,12.7,3.1,6} [sto→] VS
{600,1220,655,320,1560} [sto→] R1
{ 2100, 3900, 590, 780, 4820 } [sto→] R2
After defining the lists, type vo(vs,r1,r2) [enter] *to get the result:*

{3.89   6.09   6.02   2.20   4.53}. *This gives you the respective outputs.*

Remark: You don't need to use the same names for the lists or variables as they are shown in the definition. You can even enter the values directly, as in vo(5,600,2100), or vo(5,{600,1120},{2100,3900}).

3.1.2 Lists and your calculator

Now let us look at particular features to be considered for the use of lists in the calculator.

List variables

There are no restrictions for list variables names except those names specifically prohibited. To display a list variable on the screen enter the name on the command line.

Retrieving and displaying elements of a list:

The j-th element L_j of a list is retrieved on the command line writing the index between square brackets [ ]. For list L2={5.18, 3.98, 6.59, -0.245} above,

L2[2] enter → 3.98.

Editing the list

A list can be edited on the command line by redefining the new elements. For example, test the following two lines:

(-) 0.1 sto→ L 2 [2] enter → -0.1

L2 enter → {5.18 -0.1 6.59 -0.245}

Augmenting the list

We can increment the dimension n by 1 by defining a new element L_{n+1}. For example, with the previous list L1, by entering 8 sto→ L1[4] we get the new list L1 = {2,3,6,8}. If you need to add several items, you may use the `augment()` function described below.

3.1.3 List functions and list menus

The TI-nspire cx has several predefined list functions and operations *in the Statistic menu*. To access these function enter menu 6 3 and menu 6 4 to open the sub menus, and then you may hit the respective number or use the cursor key. It is also possible to write directly the name of th function on the command line.

The specifics of what each of the list operations in the menu accomplishes can be consulted in the guide books. I will describe below only those which are more important to us. If we need another one, I will explain it then.

Sum of Elements sum() Found in the List Math submenu, it yields the sum of elements in the list. For example

```
sum( {3,2,8, 6}) → 19.
```

It is possible to ask for a sum in a particular range. For example, from the second to the third element,

```
sum( {3,2,8, 6},2,3) → 10
```

Cumulative Sum cumsum() Found in List Operations submenu, it returns a list of the same dimension, showing the partial sum of consecutive elements. The first element in the list does not change, and the last element is the sum of all elements. For example,

```
cumsum( {3,2,8,6}) → {3, 5, 13, 19}.
```

Augment augment() Also found in the List Operations submenu. Concatenates lists. For example

```
augment({1,2}, {3,4}) → {1, 2, 3, 4}
```

3.2 Matrices

One of the most important features that our calculator has is the ability to work with matrices and the many matrix functions it contains. And, on the other hand, a very important application of matrices is in solving systems of linear equations.

It is worth then to devote several pages to talk about this application of matrices, introducing notation and also pointing out remarks of interest to us. I do not intend to enter into theoretical demonstrations, discussion of constraints and so on beyond our needs. For more detailed treatment of the theory of matrices, consult an appropriate reference.

3.3 Basic definitions and operations

Let us start with the basic definitions of interest in the applications. You may skip this section if you already have a good knowledge of the topic.

3.3.1 Some basic definitions

A matrix **A** of order or *dimension* m by n, (m × n), is a set of numbers ordered in a rectangular array of m rows and n columns. It is customary to enclose this array in parentheses, brackets, or braces. We use brackets in this book. For example, the following are matrices of order 2x2, 3x5, 2x1 and 1x3, respectively.

$$\mathbf{A} = \begin{bmatrix} 3 & 4 \\ 2 & 6 \end{bmatrix} \quad \mathbf{B} = \begin{bmatrix} 2 & -1 & 4 \\ 0 & 2 & -6 \\ 9 & 2.1 & 6 \\ 7 & 8 & -5 \\ 1 & 3 & 5 \end{bmatrix} \quad \mathbf{x} = \begin{bmatrix} 4 \\ 11 \end{bmatrix} \quad \mathbf{d} = \begin{bmatrix} 4 & 5 & 6 \end{bmatrix}$$

Matrices with only one column, like **x**, are called *column vectors*. Those with one row, like **d**, are called *row vectors*. A matrix with same number of rows and columns is a *square matrix*. **A** above is square of order 2. A *zero matrix* is one in which all elements are 0. A *unit matrix* of order n is a square matrix in which all diagonal elements are 1, and outside the diagonal are 0.

Remark: *hereafter, a column vector will be referred to simply as "vector". When necessary, distinction will be made explicitely.*

A scalar is a number, although it can also be represented by matrix of order 1x1, if you want to complicate your life. When using variables, it is customary to denote scalars by italicized or cursive, lower case letters (e.g., x), to denote vectors by bold, lower case letters as in **x**, and matrices with more than one row and one column by bold, upper case letters like **X**.

In general, to simplify notation, a matrix of order m by n can be denoted as $\mathbf{A} = [a_{ij}]_{m \times x}$. The subscript m×n may be omitted if there is no ambiguity. The notation a_{ij} represents the element in row i and column j. For vectors we may omit the number "1". In the above matrices

$$a_{12} = 2 \text{ for matrix } \mathbf{A}; \ b_{42} = 8 \text{ for matrix } \mathbf{B};$$
$$x_2 = 11 \text{ for vector } \mathbf{x}; \text{ and } d_2 = 5 \text{ for row vector } \mathbf{d}.$$

Convention *In order to be consistent with our calculator, the element in row i and column j of a matrix* **A** *will be denoted as* $\mathbf{A}[i, j]$.

3.3.2 Submatrices

A submatrix of **A** is a matrix formed by selecting from this matrix a subset of the rows and a subset of the columns of **A**, and then forming a new matrix by using those entries that appear in both the rows and columns of those selected, respecting the same order. In practice, you delete from **A** the rows and columns that are not in the submatrix.

For example, the following is a submatrix a the previous **B** formed with rows 1, 3 and 5, ad columns 1 and 3. We illustrate the process by showing the deletion of the elements:

$$\mathbf{B} = \begin{bmatrix} 2 & \cancel{1} & 4 \\ \cancel{0} & \cancel{2} & \cancel{6} \\ 9 & \cancel{21} & 6 \\ \cancel{7} & \cancel{8} & \cancel{5} \\ 1 & \cancel{3} & 5 \end{bmatrix} \Rightarrow \begin{bmatrix} 2 & 4 \\ 9 & 6 \\ 1 & 5 \end{bmatrix}$$

Notice that each column (row) of a matrix **A** is a submatrix of **A**.

3.3.3 Partitioning of matrices

A matrix **A** can be shown as an array of matrices, each one being a submatrix of **A**. For example, the following expression shows a 5x5 matrix partitioned in submatrices $\mathbf{A}_{11}$, $\mathbf{A}_{12}$, $\mathbf{A}_{21}$, and $\mathbf{A}_{22}$.

$$\mathbf{A} = \begin{bmatrix} a_{11} & a_{12} & a_{13} & a_{14} & a_{15} \\ a_{21} & a_{22} & a_{23} & a_{24} & a_{25} \\ a_{31} & a_{32} & a_{33} & a_{34} & a_{35} \\ a_{41} & a_{42} & a_{43} & a_{44} & a_{45} \\ a_{51} & a_{52} & a_{53} & a_{54} & a_{55} \end{bmatrix} = \begin{bmatrix} \mathbf{A}_{11} & \mathbf{A}_{12} \\ \mathbf{A}_{21} & \mathbf{A}_{22} \end{bmatrix} \quad (3.6)$$

where

$$\mathbf{A}_{11} = \begin{bmatrix} a_{11} & a_{12} & a_{13} \\ a_{21} & a_{22} & a_{23} \\ a_{31} & a_{32} & a_{33} \end{bmatrix} ; \mathbf{A}_{12} = \begin{bmatrix} a_{14} & a_{15} \\ a_{24} & a_{25} \\ a_{34} & a_{35} \end{bmatrix} ; \mathbf{A}_{21} = \begin{bmatrix} a_{41} & a_{42} & a_{43} \\ a_{51} & a_{52} & a_{53} \end{bmatrix}$$

and

$$\mathbf{A}_{22} = \begin{bmatrix} a_{44} & a_{45} \\ a_{54} & a_{55} \end{bmatrix}$$

Particular cases of partitions are those with the column vectors $\mathbf{a}_j$ and row vectors $\mathbf{a}^{(j)}$. If **A** is of order $m \times n$,

$$\mathbf{A} = \begin{bmatrix} \mathbf{a}_1 & \mathbf{a}_2 \cdots \mathbf{a}_n \end{bmatrix} \text{ and } \mathbf{A} = \begin{bmatrix} \mathbf{a}^{(1)} \\ \mathbf{a}^{(2)} \\ \vdots \\ \mathbf{a}^{(m)} \end{bmatrix} \quad (3.7)$$

We will use these forms for explanations later.

3.3. BASIC DEFINITIONS AND OPERATIONS

3.3.4 One remark on multiplication of matrices

Please review the matrix operations of addition, subtraction, multiplication, scalar multiplication, and transposition. I only want to bring to your attention a property of multiplication. Namely, when you multiply two matrices, we have the following property:

$$\mathbf{A}\mathbf{B} = [\mathbf{A}\,\mathbf{b}_1 \quad \mathbf{A}\,\mathbf{b}_2 \cdots \mathbf{A}\mathbf{b}_n] \tag{3.8}$$

That is, if $\mathbf{b}_j$ is the j-th column vector of $\mathbf{B}$, then $\mathbf{A}\mathbf{b}_j$ is the j-th column vector of the product $\mathbf{A} \cdot \mathbf{B}$.

3.3.5 Matrices and linear combinations

A linear combination of variables $z_1, z_2, \ldots, z_n$ is an expression of the form

$$a_1\,z_1 + a_2\,z_2 + \ldots a_n\,z_n \tag{3.9}$$

which can be expressed as a multiplication of a row vector for the coefficient a_i and a column vector for the variables z_i as

$$[a_1 \quad a_2 \cdots a_n] \begin{bmatrix} z_1 \\ z_2 \\ \vdots \\ z_m \end{bmatrix} \tag{3.10}$$

Now, for a set of linear combinations, we can express the set either as a linear combination with vector coefficients or as a matrix by a vector multiplication. This equivalence allows us to develop the background theory for practical procedures with our calculator. The development is shown next in expressions (3.11) using three variables for easy reading, but can be expanded to any number of variables without problem

$$\begin{bmatrix} a_{11}\,z_1 + a_{12}\,z_2 + a_{13}\,z_3 \\ a_{21}\,z_1 + a_{22}\,z_2 + a_{23}\,z_3 \\ a_{31}\,z_1 + a_{32}\,z_2 + a_{33}\,z_3 \end{bmatrix} = \begin{bmatrix} a_{11} \\ a_{21} \\ a_{31} \end{bmatrix} z_1 + \begin{bmatrix} a_{12} \\ a_{22} \\ a_{32} \end{bmatrix} z_2 + \begin{bmatrix} a_{13} \\ a_{23} \\ a_{33} \end{bmatrix} z_3 \tag{3.11a}$$

and

$$\begin{bmatrix} a_{11}\,z_1 + a_{12}\,z_2 + a_{13}\,z_3 \\ a_{21}\,z_1 + a_{22}\,z_2 + a_{23}\,z_3 \\ a_{31}\,z_1 + a_{32}\,z_2 + a_{33}\,z_3 \end{bmatrix} = \begin{bmatrix} a_{11} & a_{12} & a_{13} \\ a_{21} & a_{22} & a_{23} \\ a_{31} & a_{32} & a_{33} \end{bmatrix} \begin{bmatrix} z_1 \\ z_2 \\ z_3 \end{bmatrix} \tag{3.11b}$$

Now, let us bring this to a practical interpretation which in fact will become a powerful tool. The vector of linear combinations in the left is expressed in the form

$$\mathbf{a}_1 z_1 + \mathbf{a}_2 z_2 + \mathbf{a}_3 z_3 = [\mathbf{a}_1 \quad \mathbf{a}_2 \quad \mathbf{a}_3]\mathbf{z} = \mathbf{A}\mathbf{z}$$

where $\mathbf{a}_j$ is the column vector for the coefficients of z_j in the set of linear combinations, and $\mathbf{A}$ is the matrix formed with those columns. Now, if we premultiply by a matrix $\mathbf{B}$, we have

$$\begin{aligned}\mathbf{B}\begin{bmatrix} a_{11} z_1 + a_{12} z_2 + a_{13} z_3 \\ a_{21} z_1 + a_{22} z_2 + a_{23} z_3 \\ a_{31} z_1 + a_{32} z_2 + a_{33} z_3 \end{bmatrix} &= \mathbf{B}\,\mathbf{a}_1 z_1 + \mathbf{B}\,\mathbf{a}_2 z_2 + \mathbf{B}\,\mathbf{a}_3 z_3 \\ &= [\mathbf{B}\,\mathbf{a}_1 \quad \mathbf{B}\,\mathbf{a}_2 \quad \mathbf{B}\,\mathbf{a}_3]\,\mathbf{z} = \mathbf{B}\mathbf{A}\mathbf{z}\end{aligned} \quad (3.12)$$

In other words, don't worry too much about the presence of variables. The columns of $\mathbf{B}\,\mathbf{A}$ will tell you the coefficients for each variable in the set of linear combinations that results after multiplication. We will see that this is a very valuable property for us!

3.4 Matrices and the TI-nspire cx

Matrix menu You open the matrix menu with $\boxed{\text{menu}}$ 7 and then scroll the list or press the respective key number. Some of the functions in the menu will be explained later.

Matrix variables There are no restrictions for matrix variables names except those names specifically prohibited or protected.

3.4.1 Input of operations

The typical matrix and scalar operations defined in textbooks may be entered directly on the entry line, provided dimensions are compatible. These operations are :

Addition and subtraction: $\mathbf{A} + \mathbf{B}$, $\mathbf{A} - \mathbf{B}$

Matrix multiplication: $\mathbf{A} \cdot \mathbf{B}$

Multiplication or division by a scalar m: $m \cdot \mathbf{A}$, $\mathbf{A}/m$

Raise to a power for a square matrix: $\mathbf{A}^m$

Inverse of a square matrix $\mathbf{A}^{-1}$

Functions and transformations are explained later.

3.4. MATRICES AND THE TI-NSPIRE CX

Dot operations in the TI-nspire cx

Dot operations, or element operations, are allowed between matrices of equal dimensions. A dot followed by an operation key, (.<OP>), between two matrices $\mathbf{A} = [a_{ij}]$ and $\mathbf{B} = [b_{ij}]$ of similar dimensions, yields a new matrix $\mathbf{C} = [c_{ij}]$ where $c_{ij} = a_{ij}\, OP\, b_{ij}$. For example, if

$$\mathbf{A} = \begin{bmatrix} 3 & 4 \\ 2 & 6 \end{bmatrix} \qquad \mathbf{B} = \begin{bmatrix} 2 & -1 \\ 0 & 2 \end{bmatrix}$$

then

$$\mathbf{A}.*\mathbf{B} = \begin{bmatrix} 6 & -4 \\ 0 & 1 \end{bmatrix}; \qquad \mathbf{A}.\wedge\mathbf{B} = \begin{bmatrix} 4 & -0.25 \\ 1 & 36 \end{bmatrix}$$

Dot operations are also valid between a matrix and a scalar:

$$\mathbf{A}.+5 = \begin{bmatrix} 8 & 9 \\ 7 & 6 \end{bmatrix}; \qquad \mathbf{A}.\wedge 2 = \begin{bmatrix} 9 & 16 \\ 4 & 36 \end{bmatrix}$$

3.4.2 Creating matrices

Matrices in the calculator can be created using the editor tool or on the command line. Whatever method you choose, don't worry too much about matrix elements with operations involved. Just write them while entering the elements, and let the calculator do the work. If the expression is too complicated, I suggest to first work the expression and store it to a variable which you will later enter where needed.

Using the calculator editor tool

In the particular case of the TI-nspire cx, this is my preferred method. Either hit [catlg] and select any of the matrix or vector icons, or press [menu] 7 1 1. Then fill in the number of rows and columns. A void matrix will open on the command line, where you can fill up the fields. Remember, to change fields you hit [tab]

On the command line

A matrix defined on the command line must be enclosed in brackets. You can either enter each row in brackets with elements separated by commas, or separate row using semicolon (;). Both entries below create the same matrix:

[[1,2][3,4]] [enter].   [1,2 ; 3,4] [enter].

Let us now create the more complicated matrix below. I have written coefficients with explicit operations on purpose, to stress that we can enter the operations directly and let the calculator do them. Let us now solve with the calculator.

$$\begin{bmatrix} (35+126.7) & -\frac{72}{17} & 127 \\ \frac{189}{3.78} - 2.13 \times 4 & 0 & 115 \\ 3 & 44 & -11 \end{bmatrix} \qquad (3.13)$$

Separating the elements in the row with a comma (,) and the rows with semicolon (;), this matrix **A** can be entered as:

`a:= [35+126.7,` (-) `72/17,127; 189/3.78-2.13` × `4, 0, 115; 3,44,` (-) `11]`

With row vector also between brackets, the matrix would be entered as:

`a:= [[35+126.7,` (-) `72/17,127] [189/3.78-2.13` × `4,0,115] [3,44,` (-) `11]]`

If you prefer the editor, open it directly or with the ▦ key:

`a:=` menu `7 1 1 ◀ 3` tab `◀ 3` enter `35+126.7` tab `(-) 72/17` tab `127` tab `189/3.78-2.13` × `4` tab `0` tab `115` tab `3` tab `44` tab `(-) 11` tab

The result in any case, using FLOAT 5 as setting, is

$$\begin{bmatrix} 161.7 & -4.2353 & 127 \\ 41.48 & 0 & 115 \\ 3 & 44 & -11 \end{bmatrix} \qquad (3.14)$$

Notice that the commas that you enter to separate elements in rows are not displayed on output.

3.4.3 Retrieving and editing elements

To retrieve an the element at row i and column j of matrix **a** enclose the indices between brackets. Using (3.14),

`a[k,j]:   a[3,2]   → 44`

You can edit an element in a matrix directly on the entry line by storing the new value to the element. Try the following:

`5` sto→ `a[1,2]:a` enter

3.4.4 Retrieving and editing rows and columns

The row vector k of matrix **a** is denoted as `a[k]`. Using (3.14) for example,

`a[2]   → [41.48   0   115 ]`

Using the transpose function from the matrix menu it is possible to retrieve columns. However, the result may be a row vector or a column vector, depending on how you enter the command. Again, using the same matrix we

3.4. MATRICES AND THE TI-NSPIRE CX

already have see the difference between $\mathbf{a}^T$[2] and $\mathbf{a}^T$[2]^T. The transpose operation is item 2 in the matrix menu.

You can also retrieve a column vector with the use of submatrices, as explained in the next section.

To edit a complete row, simply store the new row using the same notation, as in

[-3 1 105] $\boxed{\text{sto}\to}$ a[2]: a $\boxed{\text{enter}}$ will yield, assuming that **A** is the original in (3.14),

$$\begin{bmatrix} 161.7 & -4.2353 & 127 \\ -3 & 1 & 105 \\ 3 & 44 & -11 \end{bmatrix}$$

If you want to edit a column, first transpose the matrix so you can work with the rows. For example, starting with the same matrix **A** from (3.14),

$\mathbf{a}^T$ $\boxed{\text{sto}\to}$ a: [-3 1 105] $\boxed{\text{sto}\to}$ a[2]: $\mathbf{a}^T$ $\boxed{\text{sto}\to}$ a: a $\boxed{\text{enter}}$

will yield

$$\begin{bmatrix} 161.7 & -3 & 127 \\ 41.48 & 1 & 115 \\ 3 & 105 & -11 \end{bmatrix}$$

3.4.5 Submatrices

In the TI-nspire cx, the submatrix function subMat() can only define submatrices of continuous rows and columns. This function is option 7 in the Create menu: $\boxed{\text{menu}}$ $\boxed{7}$ $\boxed{1}$ $\boxed{1}$ $\boxed{7}$.

The format for the function is subMat(matrix, j,k,l,m). Here, j,k are the row and column of the first element in the submatrix, and l, m that of the last element. If only two parameters are provided, those are of the first element, the last element being the one at the last column and row default. Using again (3.14) for our reference

subMat(a, 1,2,2,3) and subMat(a, 1,2,3,2)

yield, respectively,

$$\begin{bmatrix} -4.2353 & 127 \\ 0 & 115 \end{bmatrix} \text{ and } \begin{bmatrix} -4.2353 \\ 0 \\ 44 \end{bmatrix}$$

Observe that to retrieve the vector column j from matrix $\mathbf{A}_{m\times n}$ we write subMat(a,1,j,n,j).

3.4.6 Matrix functions in the matrix menu

When you display the matrix menu in your calculator, several functions are available. Those of interest for us are described next. The reader can consult the guidebook for a complete description of the other functions. Also, many list functions are applicable to matrices.

Transposed function T: to obtain the transposed of a matrix

Determinant function det() to find the determinant of a square matrix. In this book, this function is used mainly to check the validity of a solution process in programming. It has of course other uses.

Reduced row-echelon form function rref() of interest to us to solve systems of linear equations, as illustrated in next section.

Augment function augment() To create matrices from blocks. See subsection 3.4.6 on the facing page and the example there. You find it in the submenu Create.

Zero Matrix function newMat() Typing newMat(a,b) on the entry line produces a zero matrix of order $a \times b$. In this book, it is used mainly in initiation steps for programming. You find it in the submenu Create.

Simultaneous The function simult() to solve systems of equations. The command simult(a,b) executes $a^{-1} \cdot b$

Submatrix function submat() Already illustrated in the previous section.

Dimensions This menu item has three functions related to the size of a matrix: dim(), rowDim(), and colDim(). The first one yields a list with the number of rows and number of columns for the matrix. The others are self explanatory. They are of interest to us mainly for programming purposes.

Row operations These are the following. Notice that the result is not stored to a variable other than ANS. *Thus, if you need to work with the new matrix, store it onto a variable.* In particular, you can use the same variable, as in rowSwap(a,j,k) sto→ a.

1. rowSwap(a,j,k) yields a matrix where rows j and k in matrix a have been swapped.
2. rowAdd(a,j,k) adds row j to row k of matrix a. You can also do it with a[j]+a[k] sto→ a[k] to do it on the same matrix. .
3. mRow(m,a,j) multiplies row j of matrix a by m. Also consider m*a[j] sto→ a[k].
4. mRowAdd(m,a,j,k) multiplies row j of matrix a by m and adds the result to row k. Its equivalent would be m*a[j]+a[k] sto→ a[k].

3.4. MATRICES AND THE TI-NSPIRE CX

Column operations There are no column operations included in the menu. These can be done by first transposing the matrix, apply the row operations, and transpose back the result. For example, to swap columns 1 and 2, we could enter

a^T [sto→] a: rowSwap(a,1,2) [sto→] a: a^T [sto→] a.

We could also define column functions as follows.

1. Swapping columns: `Define colSwap(a,j,k) = (rowSwap(`a^T`,j,k))`^{T}

2. Adding columns: `Define colAdd(a,j,k) = (rowAdd(`a^T`,j,k))`^{T}

3. Multiplying a column by a constant:
`Define mCol(m,a,k) = (mRow(m,`a^T`,k))`^{T}

4. Adding m times column j to column k:
`Define mColAdd(m,a,j,k) = (mRowAdd(m,`a^T`,j,k))`^{T}

Using augment() function

The augment function is useful in many ways. Among other uses, it allows us to build a matrix using other matrices as blocks. That is, given matrices **A**, **B**, **C**, and **D**, of appropriate dimensions, it is possible to construct matrices where these ones are submatrices, of the forms

$$[\mathbf{A} \ \mathbf{B}] \quad \begin{bmatrix} \mathbf{A} \\ \mathbf{C} \end{bmatrix} \quad \begin{bmatrix} \mathbf{A} & \mathbf{B} \\ \mathbf{C} & \mathbf{D} \end{bmatrix}$$

The command `augment(a,b)` will produce the first $[\mathbf{A} \ \mathbf{B}]$.

With a semicolon, `augment(a;c)` produces the second form.

More complex forms, like the third one shown above, can be built by steps, as illustrated with the following example.

Example 3.5 *Assume we have already the following matrices, stored in* a, b, c, d

$$\mathbf{A} = \begin{bmatrix} 1 & 2 \\ 3 & 4 \end{bmatrix}; \quad \mathbf{B} = \begin{bmatrix} -1 & -2 \\ -3 & -4 \end{bmatrix}; \quad \mathbf{C} = \begin{bmatrix} 5 & 6 \end{bmatrix}; \quad \mathbf{D} = \begin{bmatrix} 9 & 10 \end{bmatrix}$$

Then

`augment(a,b)` [enter] → $\begin{bmatrix} 1 & 2 & -1 & -2 \\ 3 & 4 & -3 & -4 \end{bmatrix}$

`augment(a;c)` [enter] → $\begin{bmatrix} 1 & 2 \\ 3 & 4 \\ 5 & 6 \end{bmatrix}$

Let us use the colon (:) to enter several commands with one line next. You can build now the matrix:

augment(a;c) `sto→` f: augment(b;d) `sto→` g:

augment(f,g) `enter` → $\begin{bmatrix} 1 & 2 & -1 & -2 \\ 3 & 4 & -3 & -4 \\ 5 & 6 & 9 & 10 \end{bmatrix}$

You can also do it in one step as augment(augment(a;c),augment(b;d))

3.5 Matrices and Linear equations

Matrices and systems of linear equations are intimately associated since ancient times. To simplify discussion and space, I work here with 3 variables; generalization is straightforward. For now, the objective is to introduce notation and talk about dealing with matrices with the calculator.

$$\begin{aligned} a_{11}x_1 + a_{12}x_2 + a_{13}x_3 &= b_1 \\ a_{21}x_1 + a_{22}x_2 + a_{23}x_3 &= b_2 \\ a_{31}x_1 + a_{32}x_2 + a_{33}x_3 &= b_3 \end{aligned} \qquad (3.15)$$

which can be expressed in matrix equation as

$$\begin{bmatrix} a_{11} & a_{12} & a_{13} \\ a_{21} & a_{22} & a_{23} \\ a_{31} & a_{32} & a_{33} \end{bmatrix} \begin{bmatrix} x_1 \\ x_2 \\ x_3 \end{bmatrix} = \begin{bmatrix} b_1 \\ b_2 \\ b_3 \end{bmatrix} \qquad (3.16)$$

The 3x3 matrix is the coefficient matrix **A**, formed with the coefficients of the variables. Observe that the rows in this matrix are the coefficients in the equations, while the columns are the coefficients of variables in different equations. Using labels, we illustrate this as follows

$$\mathbf{A} = \begin{matrix} \\ \text{Eq. 1} \\ \text{Eq. 2} \\ \text{Eq. 3} \end{matrix} \begin{matrix} x_1 & x_2 & x_3 \\ \begin{bmatrix} a_{11} & a_{12} & a_{13} \\ a_{21} & a_{22} & a_{23} \\ a_{31} & a_{32} & a_{33} \end{bmatrix} \end{matrix} \qquad (3.17)$$

I will use labels for rows and columns to explain the structure of the matrices, simplify explanations and so on.

On the right hand side of (3.16) we find the vector of knowns. Instead of a vector, it could be a matrix, as illustrated in subsection 3.5.2. This vector or matrix of knowns can be generated for different reasons and depending on the application.

3.5.1 Solving linear equations

To solve system (3.16) in your TI-nspire cx, create matrices **A** and **b**. We have then the alternatives mentioned below. Remember that the solution exists only and only if the determinant of **A** is not zero.

(A) Use the embedded function Simultaneous, simult(a,b), which is called with `menu` 7 6.

(B) Enter on the command line the sequence A `∧` `(-)` 1 `×` b.

(C)) Generate the augmented matrix **C** = [**A** **b**] and apply the reduced row echelon form transformation (**rref(C)**).

In the first two methods, the answer is the desired solution. In the last one, the solution is read from the last column. These three methods are illustrated with the next example.

Example 3.6 *Solve the following set of equations:*

$$
\begin{aligned}
(35 + 126.7)\,x_1 - \tfrac{72}{17} x_2 + 127\,x_3 &= 109 \\
\left(\tfrac{189}{3.78} - 2.13 \times 4\right) x_1 + 0\,x_2 + 115\,x_3 &= -87 \\
3\,x_1 + 44\,x_2 - 11\,x_3 &= -7
\end{aligned}
$$

The coefficient matrix is the same as in expression (3.13) on page 34. The vector of knowns **B** *may be similarly created. Now you can do the following:*

(A) Press `menu` 7 6 a b ⇒ simult(a,b) `enter` or

(B) Press a `∧` `(-)` 1 `×` B `enter`
Both will produce the vector

$$
\begin{bmatrix} 1.74673 \\ -.634827 \\ -1.38656 \end{bmatrix}
$$

which is interpreted as $x_1 = 1.74673$, $x_2 = $ -634827, and $x_3 = $ -1.38656.

(C) On the other hand, using **rref(** together with the **augment(** transformation:

rref(augment(A,B)) `enter`

yields

$$\begin{bmatrix} 1 & 0 & 0 & 1.74673 \\ 0 & 1 & 0 & -.634827 \\ 0 & 0 & 1 & -1.38656 \end{bmatrix}$$

The last column is the solution.

3.5.2 Multiple systems and knowns.

There are several situations in which a matrix **B** is used instead of vector **b** in the above process. The interpretation of the results depends on the application. I illustrate with two examples.

Example 3.7 *Take two systems in which the only difference is the knowns-vector:*

$$\begin{array}{ll} 3x + 2y = 8 & \qquad 3x + 2y = 1 \\ -x + 3y = 1 & \qquad -x + 3y = 7 \end{array}$$

Now create the coefficient matrix **A** *for the equations, and the matrix* **B** *using the vectors as columns:*

$$\mathbf{A} = \begin{bmatrix} 3 & 2 \\ -1 & 3 \end{bmatrix}$$

and

$$\mathbf{B} = \begin{bmatrix} 8 & 1 \\ 1 & 7 \end{bmatrix}$$

Then

$$\mathbf{A}^{-1}\mathbf{B} = \begin{bmatrix} 3 & 2 \\ -1 & 3 \end{bmatrix}^{-1} \times \begin{bmatrix} 8 & 1 \\ 1 & 7 \end{bmatrix} = \begin{bmatrix} 2 & -1 \\ 1 & 2 \end{bmatrix} \qquad (3.18)$$

The same matrix is obtained with simult(A,B).
Reading the result: For the first set, $x = 2$ and $y = 1$. For the second set: $x = -1$ and $y = 2$.
On the other hand, the entry `rref(augment(A B))` `enter` yields

$$\begin{bmatrix} 1 & 0 & 2 & -1 \\ 0 & 1 & 1 & 2 \end{bmatrix}$$

and we now read the solutions from the submatrix at the right of the unit submatrix.

Let us look at another situation with the same matrices:

3.5. MATRICES AND LINEAR EQUATIONS

Example 3.8 *Take the system*

$$3x + 2y = 8 + z$$
$$-x + 3y = 1 + 7z$$

We see that the right hand side of the system is a set of linear combinations. That means that the system can be expressed as

$$\mathbf{A}\,\mathbf{x} = \mathbf{B} \begin{bmatrix} 1 \\ z \end{bmatrix}$$

where $\mathbf{A}$ *and* $\mathbf{B}$ *are the same matrices of the previous example.*

Using the property illustrated by expression (3.12) on page 32, we can work with simult($\mathbf{A}$,$\mathbf{B}$) *or multiply* $\mathbf{A}^{-1}\mathbf{B}$. *In both cases, we get the result (3.18).*

The important step here is interpretation. Although the matrix result now is the same as in the previous example, the situation is different and now we see the solution for each variable from the rows: The first row means x = 2 - z, *and the second row means* y = 1 + 2z.

Notice how important is the fact that it is not only a matter of entering data onto the calculator and getting an answer. You must interpret your results!

3.5.3 Exchanging knowns and unknowns

In subsection 3.3.4 I mentioned the fact that the multiplication of a matrix $\mathbf{A}$ and a column vector $\mathbf{x}$ can be expressed as a linear combination

$$\mathbf{a}_{(1)}\, x_1 + \mathbf{a}_{(2)}\, x_2 + \ldots \mathbf{a}_{(n)}\, x_n$$

where $\mathbf{a}_{(m)}$ is the m-th column of the coefficient matrix $\mathbf{A}$ and $x_1, x_2, \ldots$ are the elements of vector $\mathbf{x}$. Let us now use this property to illustrate a very useful procedure that allows us to exchange "knowns and unknowns" in a problem expressed as a set of equations as illustrated next.

The system of equations in example 3.8 can also be expressed as

$$\mathbf{a}_{(1)}\, x + \mathbf{a}_{(2)}\, y = \mathbf{b}_{(1)} + \mathbf{b}_{(2)}\, z \qquad (3.19)$$

In this system it is assumed that variable z is known. In practical terms, we say that we want to express x and y as functions of z. Now, imagine that there is a situation in another problem where where x is known and z unknown, and yet the relational equations are the same as before and you already have the matrices $\mathbf{A}$ and $\mathbf{B}$ available. How can we proceed?

One way is to rewrite the new set of equations. This is equivalent to exchange, with proper sign modifications, the respective columns from both matrices so we can represent the new problem as

$$-\mathbf{b}_{(1)} z + \mathbf{a}_{(2)} y = \mathbf{b}_{(1)} - \mathbf{a}_{(2)} x$$

Let us illustrate the whole process with the numerical example given before.

Example 3.9 *The original set is*

$$\begin{aligned} 3x + 2y &= 8 + z \\ -x + 3y &= 1 + 7z \end{aligned}$$

and what we want to solve now is

$$\begin{aligned} -z + 2y &= 8 - 3x \\ -7z + 3y &= 1 + x \end{aligned}$$

We already have the matrices **A** and **B** from the original set:

$$\mathbf{a} = \begin{bmatrix} 3 & 2 \\ -1 & 3 \end{bmatrix} \text{ and } \mathbf{b} = \begin{bmatrix} 8 & 1 \\ 1 & 7 \end{bmatrix}$$

Our objective is to manipulate these matrices so we can solve the new set. I show the steps below, omitting the enter *key at the end of each command.*

Remember that the calculator does not work with or access columns. Since it can work with rows, we must first transpose our matrices and work with the corresponding rows. Then we transpose again to get the desired result. Let us first use the embedded functions in our calcullator. After that, let us make the changes ourselves by retrieving and editing the rows.

Step 1: Transpose and store the augmented matrix C=[A,B]^T
(augment(a,b))^T sto→ c

Step 2: Exchange rows 1 and 4 of C *This steps takes care of exchanging columns 1 and 4 from the original augmented matrix* **C**:
rowSwap(c,1,4) sto→ c

Step 3: Multiply row 1 by -1, and repeat with row 4 :
mrow((-) 1,c,1) sto→ c:mrow((-) 1,c,1) sto→ c

Step 4: Transpose [C *again]: This will yield the desired set of equations in augment form:* c^T sto→ c

Step 5: Apply RREF function to solve rref(c) enter

The last step results in the matrix

$$\begin{bmatrix} 1 & 0 & 2 & -1 \\ 0 & 1 & 5 & -2 \end{bmatrix}$$

Hence, $z = 2 - x$ and $y = 5 - 2x$.

Let us now work it by retrieving and editing the rows, using an auxiliary variable or the stack.

3.5. MATRICES AND LINEAR EQUATIONS 43

Step 1: transpose: a^T ⌊sto→⌋ a : b^T ⌊sto→⌋ b

Step 2: retrieve "column" 1 of **A** and store: a[1] ⌊sto→⌋ x

Step 3: Exchange "columns" with sign changes:
⌊(-)⌋ b[2] ⌊sto→⌋ a[1]:   ⌊(-)⌋ x ⌊sto→⌋ b[2]

Step 4: transpose again: a^T ⌊sto→⌋ a : b^T ⌊sto→⌋ b

Step 5: solve a^{-1} * b *or* simult(a,b)

The example chosen was simple enough for the reader to check by hand calculations. This procedure, however, becomes very handy in many situations, as we will see in later chapters.

3.5.4 Closing remarks

A relative brief visit to lists and matrices was the goal of this chapter, considering only those characteristics and operations of interest for the goals of this book. The topic can be extended to the point of making a book just for them.

You will notice all along the book that I use the notation a^{-1} b as well as a ⌊(-)⌋ 1 × b instead of the embedded function simult(a,b). This notation follows because we focus in learning, and repeating now and then how you can do things directly is worth the writing.

We are now ready to go into circuit analysis.

CHAPTER 4

Four Network Theorems and Applications

You may skip this chapter and come back later if you need refreshing. It is of theoretical nature, but very practical consequences. The theorems presented in this chapter provide the theoretical background for many applications of our calculator. Remark that the applications themselves are doable with pencil and paper, but that's another story.

There are four very important theorems for circuits: a) the *Voltage Source substitution theorem*, the *Current Source Substitution Theorem*, applicable to all circuits, and b) the *Homogeneity Theorem* and the *Superposition Theorem* for linear circuits. In this chapter I also use the first three theorems to illustrate how we can combine theory with calculator, and obtain results for situations in the same circuit with different sources or information. A full chapter is devoted later to superposition theorem.

4.1 Substitution Theorems

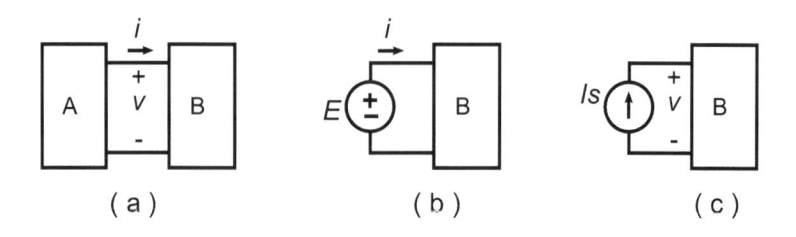

Figure 4.1: Substitution Theorems: (a) Case; (b) Voltage Substitution; (c) Current Substitution

4.1. SUBSTITUTION THEOREMS

The substitution theorems state the following:

Voltage Source Substitution Theorem: Let a circuit be partitioned in two subcircuits A and B as shown in Fig. 4.1(a). If the voltage v is known, say $v = E$, than subcircuit A may be substituted by a voltage source of value E without any change in currents and voltages of subcircuit B, including the current i entering the subcircuit. This is illustrated in Fig. 4.1(b)

Current Source Substitution Theorem: Let a circuit be partitioned in two subcircuits A and B as shown in Fig. 4.1(a). If the current i is known, say $i = Is$, than subcircuit A may be substituted by a current source of value Is without any change in currents and voltages of subcircuit B, including the voltage v at the terminals of B. This is illustrated in Fig. 4.1(c)

These theorems are valid provided no mathematical inconsistency is introduced. Let us start with an example

Example 4.1 *An analysis of the circuit in Fig. 4.2(a) yields the following results:*

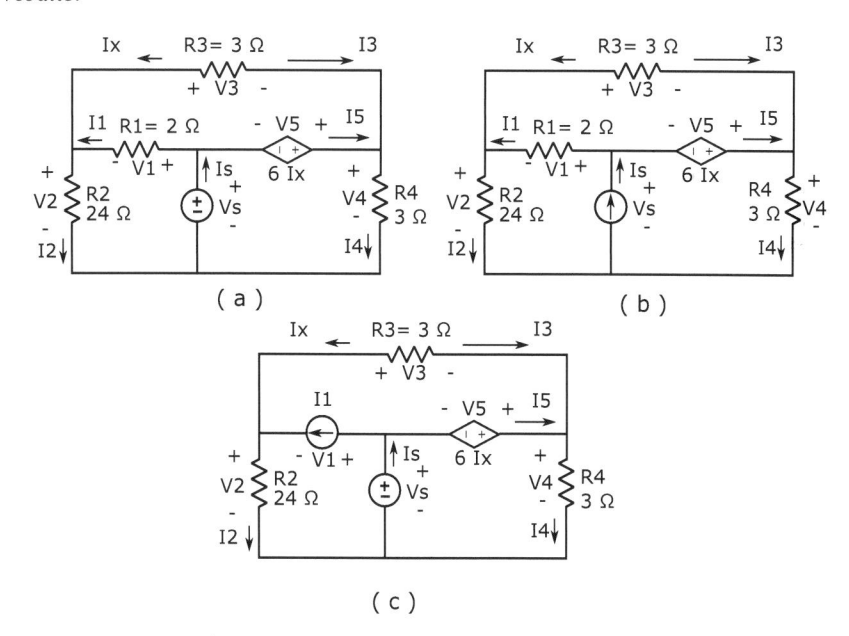

Figure 4.2: Example of Substitution Theorems: (a) Original circuit; (b) Voltage source substituted by current source; (c) Current Substitution of resistor R1

46 CHAPTER 4. FOUR NETWORK THEOREMS AND APPLICATIONS

Vs = 30 V	Is = 7 A	V1 = 6 V	I1 = 3 A
V2 = 24 V	I2 = 1 A	V3 = 6 V	I3 = 2 A (= - Ix)
V4 = 18 V	I4 = 6 A	V5 = - 12 V (6Ix)	I5 = 4 A

In inset (b), the voltage source has been substituted by a current source. A current source is also used in Fig. 4.2(c) substituting a resistance. The reader can verify that the voltages and currents are the same in all cases.

Remark: As it can be appreciated in the previous example, when going from circuit (a) to circuit (b), it is actually not important from the numerical point of view whether you use a voltage or a current source.

A more interesting application of the substitution theorem can be seen in the next example.

Example 4.2 *Consider the configuration in Fig. 4.3(a), where block A is a resistive linear subnetwork and the black box element is of any type, like for a example a capacitance or a non linear element.*

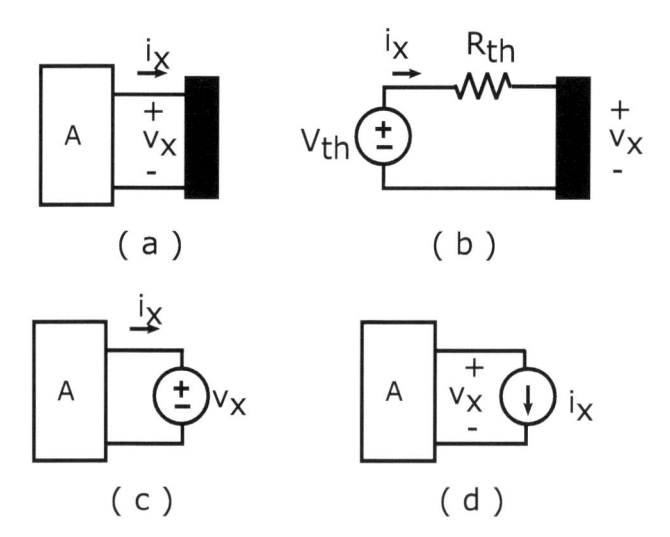

Figure 4.3: Another Example of Substitution Theorems: (a) Original circuit; (b) Block A substituted by its Thevenin's equivalent; (c) and (d) Substitution theorems applied

Fig. 4.3(b) shows the substitution of block A by its Thevenin's equivalent. This configuration is much easier to dealt with, since it reduces to two equations:

$$\text{Loop equation:} \quad V_{th} = R_{th} i_x + v_x$$

and

$$\text{Element's equation:} \quad f(v_x, i_x) = 0$$

4.2. HOMOGENEITY AND PROPORTIONALITY

The latter may be non linear, or may involve differentials, as it is the case of capacitances and inductances. Whatever the case, it is possible to obtain a solution by some means, including the use of calculator, and find the values for v_x and i_x.

Once these values are found, the black box may be substituted by either a voltage or a control source in the original circuit, as shown in Fig. 4.3(c) and (d), respectively. With this circuit we solve for the elements in block A. This is a very common way to proceed.

4.2 Homogeneity and Proportionality

Consider a circuit with only one independent source, as illustrated by Fig. 4.4. The source may be of any type. If it is a voltage source, then v_s is known and i_s is to be calculated, and vice versa. In fact, except for situations where a mathematical inconsistency may arise, the treatment of these magnitudes is a matter of mathematical convenience, as pointed out in the previous section. For general discussion, we shall speak of a source z, without any particular reference to the type of magnitude, except if confusion may arise.

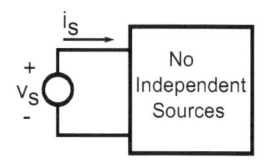

Figure 4.4: One Source Circuit

The circuits we are dealing with are linear. Because of this, the next theorem applies to them.

Theorem of Homogeneity: *In a linear network with only one independent source of value z, (Fig. 4.4), when this source is substituted by one of value Kz, where K may be a constant, then every every current and voltage in the circuit will be multiplied by K.*

Mathematically, this is usually stated in linear systems, considering y an output and z an input, as follows:

If the system is linear and $y = f(z)$, then $y_1 = f(Kz) = k\,f(z) = ky$ \quad (4.1)

REMARK: If the circuit in Fig. 4.4 contains only linear resistors, linear dependent sources, and operational amplifiers, then the principle of homogeneity is also valid when K is a time function. This special case is not applicable to circuits containing elements that are either time dependent or whose characteristic may involve time dependence, as it happens with capacitances or inductances.

For linear resistive circuits, we also have the Principle of Proportionality, from which that of Homogeneity is a consequence. This principle can be stated as follows:

Principle of Proportionality: *If a linear resistive network has only one independent source z, then every current and voltage in the circuit is of the form Kz, where K is a constant which depends exclusively on the circuit elements, but not on the source.*

Let us look at an example.

Example 4.3 *Consider the circuit from example 4.1, Fig. 4.3(a). If the source V_s is now $24e^{-100t}$ V, then every current and voltage should be multiplied by $0.8e^{-100t}$, which is the factor needed to convert the original 30 V source to the new one. The result is then*

$$\begin{array}{ll} Vs = 24e^{-100t} \text{ V} & Is = 5.6e^{-100t} \text{ A} \\ V1 = 4.8e^{-100t} \text{ V} & I1 = 2.4e^{-100t} \text{ A} \\ V2 = 9.6e^{-100t} \text{ V} & I2 = 0.8e^{-100t} \text{ A} \\ V3 = 4.8e^{-100t} \text{ V} & I3 = 1.6e^{-100t} \text{ A } (= \text{ - Ix}) \\ V4 = 13.6e^{-100t} \text{ V} & I4 = 4.8e^{-100t} \text{ A} \\ (6Ix) \; V5 = \text{ - } 9.6e^{-100t} \text{ V} & I5 = 3.2e^{-100t} \text{ A} \end{array}$$

Furthermore, if we want a specific voltage, say V_4 to be of another value A, then we multiply all elements, including the source, by $A/6$, so we have the required source value as well as the new currents and voltage.

We can, in general take a matrix or a list to have an easier way to calculate. Take for example

$$\mathbf{M} = \begin{bmatrix} 30 & 7 & 6 & 3 \\ 12 & 1 & 6 & 2 \\ 18 & 6 & -12 & 4 \end{bmatrix}$$

to represent the values of the original circuit. Changes are done in one step with the operation $K\mathbf{M}$

4.3 Superposition Theorem

This theorem is valid for all linear systems, where it is call additivity property. In our circuits, it is applied when several independent sources are present, as illustrated by Figure 4.5.

The principle can be stated as follows:

4.3. SUPERPOSITION THEOREM

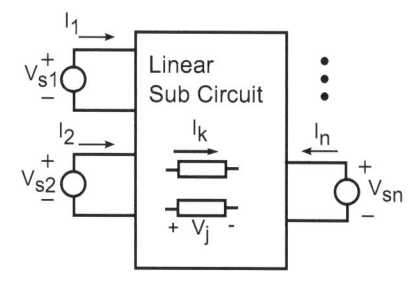

Figure 4.5: Many sources circuit

Superposition: Let a circuit have n independent sources $z_1, z_2, \ldots, z_n$. Any voltage V_j or current I_k in the circuit, including voltage or currents in the sources may be expressed in the form

$$V_j = V_{j1} + V_{j2} + \ldots + V_{jn} \qquad (4.2)$$

and

$$I_j = I_{j1} + I_{j2} + \ldots + I_{jn} \qquad (4.3)$$

where V_{jh} is the value of voltage V_j and I_{kh} the value current I_k when all independent sources except z_h, $h = 1, 2, \ldots, n$ are 0 (turned-off)

Fig. 4.6 illustrates the superposition theorem by steps. Although the theorem is valid for any linear subcircuit, the restriction to resistive circuits is enhanced to illustrate the principle below of interest to us.

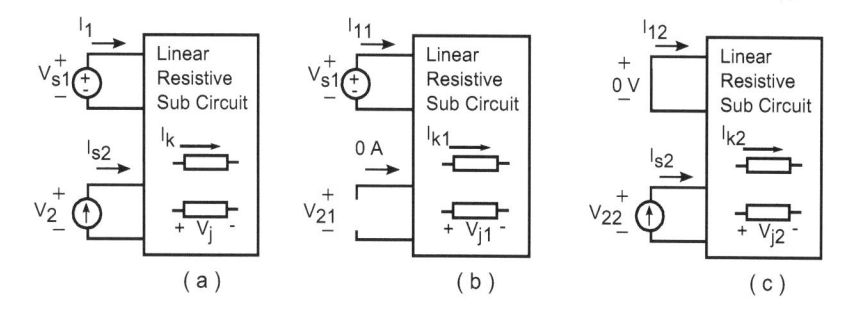

Figure 4.6: Many sources circuit

Notice that the individual calculations illustrated in steps (b) and (c) work with only one source. Hence, the properties mentioned in section 4.2 apply. In particular, we can extend the theorem of homogeneity (and proportionality) as follows:

50 CHAPTER 4. FOUR NETWORK THEOREMS AND APPLICATIONS

Let x_j be a current or a voltage in a linear resistive circuit with sources $z_1, z_2, \ldots, z_m$, with

$$x_j = x_{j1} + x_{j2} + \ldots + x_{jm}$$

where x_{jh} is the value that results when all sources except z_h are off. Then, for the set of values $\{k_1 z_1, k_2 z_2, \ldots k_m z_m\}$ the response will be

$$x_j = k_1 x_{j1} + k_2 x_{j2} + \ldots + k_m x_{jm} \tag{4.4}$$

In particular, for resistive linear networks, k_j may be a function of time.

Example 4.4 *A circuit has three sources, V1=2 V, I2 = 4 mA and V3 = 1.8 V. The response of interest is a current i_o. The following results are known:*
- *When I2 and V3 are turned off, i_o = 2.8 mA.*
- *When V1 and V3 are turned off, i_o = 4.7 mA.*
- *When V1 and I2 are turned off, i_o = 6.5 mA.*

1. *What is the value of i_o?*

2. *How do you express i_o as a function of the three sources?*

3. *What ar the individual contributions of each source to i_o when V1=3.6 V, I2 = 11.3 mA and V3 = 7.3 V? What is the the total value of i_o in this case?*

Solution: Let us set engineering display The first question can be answered by simply adding the individual values. Yet, to explore further the different questions, let us create two lists. One is for the original sources values and the other for the respective responses in i_o:

For sources V1, I2 and V3, in that order: Y:= {2, 4.⎡EE⎤ ⎡(-)⎤3, 1.8}

For i_o: {2.8 ⎡EE⎤ ⎡(-)⎤3, 4.7 ⎡EE⎤ ⎡(-)⎤3, 6.5 ⎡EE⎤ ⎡(-)⎤3} ⎡sto→⎤ X

Let us now look at the answers:

1. Since i_o is the sum of partial answers:
   sum(X) → 14.E-3 tells us that i_0 =14 mA

2. The question is answered by assuming that all sources are equal to 1, and scaling the individual contributions. That is, divide each contribution by the respective source value:

   X/Y → {1.4E-3, 1.175E0, 3.611E-3}

4.3. SUPERPOSITION THEOREM

which means

$$i_o = 1.4 \times 10^{-3} V_1 + 1.175 I_2 + 3.611 \times 10^{-3} V_3$$

3. *The first question of this item can be answered by scaling i_o contributions to the new source values. This is similar to substitute the new values in the expression of the previous item:*

{3.6,11.3 `EE` `(-)` 3,7.3} × X/Y
   → {5.04E-3, 13.278E-3, 26.361E-3}

The total value of i_o is found with

`sum(ans)` → `44.68E-3`

and therefore $i_o = 44.68$ mA

The superposition theorem will be applied to yield faster results in several applications such as calculating Thevenin and Norton equivalent circuits, two-port and multiport parameters and so on. This is why a whole chapter is devoted to this property.

When combined with other theorems, like the substitution theorem, it provides further tools to deal with circuits containing reactive elements in time domain, non linear elements, etc. These applications fall outside the scope of this book but will be introduced in a second volume.

Writing equations for superposition

Lists and matrices make it easier to work superposition in few steps. This has already been illustrated with the previous example. Other examples may be found in section 5.3.1 on page 73. Let us introduce the general way of how we set up equations for methods solving equations with matrices. This is done next using the concepts introduced in section 3.5.2 on page 40.

It can be shown that any equation of the circuit can be written as

$$a_{i1} x_1 + a_{i2} x_2 + + \ldots + a_{im} x_m = b_{i1} z_1 + b_{i2} z_2 + + \ldots + b_{in} z_n$$

All equations can be written then in matrix form as

$$\mathbf{A}\,\mathbf{x} = \mathbf{B}_{(1)} z_1 + \mathbf{B}_{(2)} z_2 + \ldots + \mathbf{B}_{(n)} z_n = \mathbf{B}\,\mathbf{z} \tag{4.5}$$

where $\mathbf{A}$ is the coefficient matrix of order $m \times m$, $\mathbf{B}_{(j)}$ is a column vector of order $m \times 1$.

Notice that if all z's are zero, except z_h, then (4.5) reduces to

$$\mathbf{A}\,\mathbf{x}_h = \mathbf{B}_{(h)} z_h \tag{4.6}$$

From the practical point of view of numerical calculations, we see that solving (4.5) we have

$$\mathbf{x} = \mathbf{A}^{-1}\mathbf{B}\,\mathbf{z} \qquad (4.7)$$

In this expression, $\mathbf{A}^{-1}\mathbf{B}$ has all the information concerning the individual contributions of the sources and thus what we need to obtain the desired results is there. The j-th column shows the contribution of source z_j.

Hence, when writing down equations, instead of putting all terms in one column, write the set in the form (4.7). In your calculator all you need is $\mathbf{B}$, with the vector $\mathbf{z}$ used for you to interpret results. Different examples of application are illustrated along the book.

4.4 Closing remarks

There are of course more theorems that those presented in this chapter, and useful to develop methods for working with calculators. Those in this chapter were specifically mentioned because they are applied throughout the entire book.

As the reader looks at examples of applications of theorems, awareness of the importance of understanding and using theorems will become more evident. Not only because of the easiness they may introduce in setting up algorithms for solution, but also because they provide us with the necessary tools to understand and broaden the meaning of results.

CHAPTER 5

Transformations and reductions

The goal of this chapter is to review and provide examples for several formulas associated to subcircuits used in reduction and transformation of circuits. In this group we have the series and parallel connections, voltage and current dividers, operational amplifiers structures and so on.

Since we are using calculators, we include formulas with resistances and conductances. Remember that if the datum is a resistance R, the conductance is applied as $1/R$ or R^{-1}.

The methods in this chapter rely on intuition, and are indeed very important for developing insight into design procedures. Of course, alternative ways to solve a problem are always possible. The solutions and methods presented try to emphasize the use of calculator or to highlight a particular feature of the calculator. They are applicable to ac circuits with complex numbers.

5.1 Series and Parallel Connections

The most basic reduction formulas are those for the equivalent resistance of series and parallel sub circuits (Figure 5.1). Let us start with them.

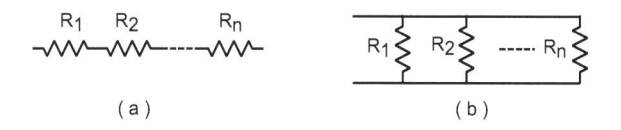

Figure 5.1: (a) Series connection of Resistances; (b) Parallel Connection of Resistances

5.1.1 Equivalent resistance and conductance formulas

$$\text{For Series Connection: } R_{eq} = R_1 + R_2 + \ldots + R_n = \sum_{j=1}^{n} R_j \qquad (5.1\text{a})$$

$$\text{For Parallel Connection: } R_{eq} = \left(\frac{1}{R_1} + \frac{1}{R_2} + \ldots + \frac{1}{R_n}\right)^{-1} \qquad (5.1\text{b})$$

The equivalent conductances in terms of resistances follow directly too. Remember that $G_{eq} = 1/R_{eq}$.

$$\text{For Series Connection: } G_{eq} = \frac{1}{R_1 + R_2 + \ldots + R_n} \qquad (5.2\text{a})$$

$$\text{For Parallel Connection: } G_{eq} = \frac{1}{R_1} + \frac{1}{R_2} + \ldots + \frac{1}{R_n} = \sum_{j=1}^{n} \frac{1}{R_j} \qquad (5.2\text{b})$$

For the particular case of two parallel resistances, (5.1b) can be reduced to the popular formula

$$\text{For Parallel Connection of two resistances: } R_{eq} = \frac{R_1 R_2}{R_1 + R_2} \qquad (5.3)$$

This form is preferred by most students for two reasons. First, data is usually given in resistance values. Second, this form involves only one division, while the deployed one involves three. For hand analysis, these are good reasons. But when working with calculators, this should not be quite a problem. Let us illustrate this point with an example.

Example 5.1 *Fig. 5.2 shows four connections of different complexity. Let us use the above formulas to find the respective equivalent resistances.*

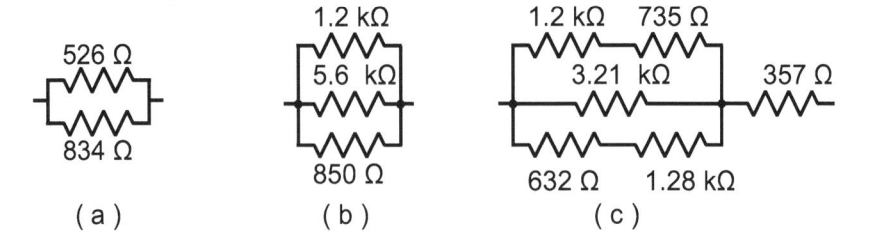

Figure 5.2: Some exampes for Series and Parallel connections

Let us work each case with separately. For this first example, I will show in particular the sequence of keys to be pressed. Later, I will limit myself to

5.1. SERIES AND PARALLEL CONNECTIONS

show mainly what appears on the command line. Remember that any expression of the form 1/N (N a number) is entered as 1 $\boxed{\div}$ N, but can also be entered as N $\boxed{\wedge}$ $\boxed{(-)}$ 1.

CASE a) Resistances 526 Ω and 834 Ω are in parallel (526||834)[1]. We find the equivalent using (5.1b) and enter

(1/526+1/834)∧ $\boxed{(-)}$ 1 $\boxed{\text{enter}}$ → 322.56,   *pressing 17 keys.*

Notice that we could have entered

1/(1/526+1/834) $\boxed{\text{enter}}$ → 322.56   *for 15 keys.*

On the other hand, (5.3) requires 19 keystrokes, which is comparable. Let's see now the other cases.

CASE b) We calculate now 1.2 kΩ||5.6 kΩ||850 Ω. For this example, I will use again (5.1b) which requires about half the time than successive applications of (5.3) when using the calculator –25 keystrokes against 40.–

1/(1/1.2E3+1/5.6E3+1/850) → 456.96

Remarks: Remember that the E symbol expresses a power of 10, and is entered with the $\boxed{\text{EE}}$ key.

Expression (5.1b) has an advantage: it allows us to also see the value of the conductance. To show this, we recalculate the equivalent resistance using an intermediate step without changing the number of keystrokes (except for an additional $\boxed{\text{enter}}$):

1/1.2E3+1/5.6E3+1/850 → 0.00219

Thus, the conductance is 2.19 mS. To get the resistance, now press $\boxed{\wedge}$ $\boxed{(-)}$ $\boxed{1}$ to automatically introduce introduce the ans variable, and the result is

ans^{-1} → 456.96

Reminder: If you start your command line hitting an operation key, the previous result is automatically entered on the entry line as **ans**.

CASE c) Now for the series connection of 357Ω with the parallel subcircuit

---

[1] We write R1 || R2|| R3 ||... to denote a parallel connection of resistances R1, R2, R3, ...

$(1.2\text{k}\Omega+735\Omega)||3.21\text{k}\Omega||(632\Omega+1.28\text{ k}\Omega)$, *we first calculate the conductance of the parallel connection. To reduce the risk of false keystrokes, it is better to do it by steps, using the property of* ans *mentioned in the remark. In addition, let us work in Engineering floating 5 modes. Do:*

```
1/(1.2E3+735)+1/3.21E3+1/(632+1.28E3)  → 1.3513E⁻3
ans ∧⁻1 + 357  → 1.097E3  or  ans⁻¹ + 357  → 1.097E3
```

Hence, the equivalent resistance is 1.1 kΩ, *rounding to one digit in the decimal fraction.*

Scaling units Since you are the one using the calculator, you may accelerate your work if you adopt the habit of using the appropriate units. For example, by adopting kilo ohms instead of ohms, you can save keystrokes. Your results for resistances are interpreted in kilo ohms while those of conductances as milli siemens. Thus the two last lines would be entered as

```
1/(1.2 + .735)+1/3.21+1/(.632+1.28)  → 1.3513 (mS)
ans ∧⁻1 + .357  → 1.097 (kΩ)
```

5.1.2 Another example of Series-Parallel Reduction

Let us look at an example which is very "hand-like", in the sense that we follow more or less the same steps as those in a hand analysis, except perhaps for the use of the operations. This type of examples appears in the many situations such as when step by step considerations are required or when a compact plan is not so direct.

Example 5.2 *Let us find the equivalent resistance of the circuit on Figure 5.3 by reducing it to only one resistance, namely, the equivalent resistance. We present two decimal figures in results.*

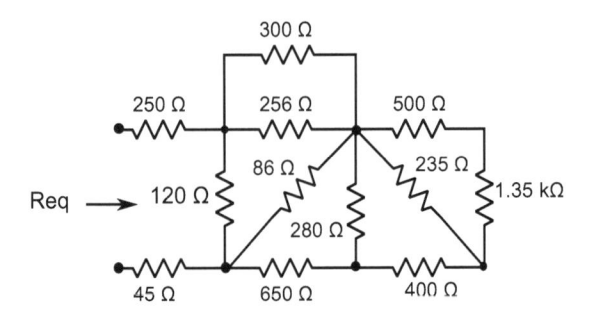

Figure 5.3: Example for series-parallel reduction

5.1. SERIES AND PARALLEL CONNECTIONS

We first identify at the right side of the circuit 400 Ω in series with the parallel connection of 235 Ω and the series connection 500 Ω- 1350 Ω. Using the formula for parallel connection, while including the series formula within the same expression, we have

$$400 + \left(\frac{1}{235} + \frac{1}{500+1350}\right)^{-1} = 608.51$$

which is entered as

400 + (1÷ 235 + 1÷ (500 + 1350)) ∧ (-) 1 enter → 608.51

The process is shown in Figure 5.4. In this figure I highlight the fact that the result is now stored onto ans.

Remember! In hand analysis you would probably have preferred the classical R1R2/(R1+R2) reduction, which is correct. But now we take advantage of calculator for the numerical steps and can introduce inverses with no remorse or fear.

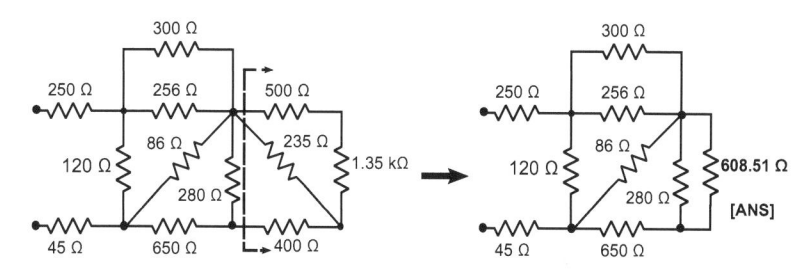

Figure 5.4: Example for series-parallel reduction, first steps

The next reduction will be done in short steps. We reduce the sub circuit in parallel with the 86 Ω resistance. This is built with a 650 Ω resistance in series with the parallel connection of 608.51 Ω –which is already on stack, as illustrated –, and 280 Ω. The reduction is done with the expression below, and the result is illustrated in Fig. 5.5.

$$650 + \left(\frac{1}{ans} + \frac{1}{280}\right)^{-1} = 841.76$$

which in two short steps can be entered as follows:

1 ÷ ans + 1 ÷ 280 enter → 5.21E-3
∧ (-) 1 + 650 enter → 841.76

We now reduce separately the parallel connections 841.76 || 86 Ω and 300 || 256 Ω. We store the intermediate step to a variable, say X:

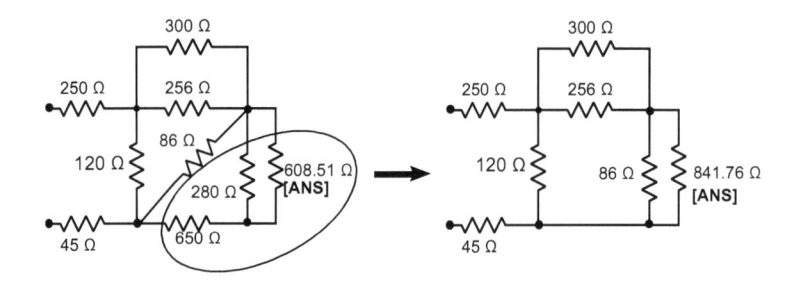

Figure 5.5: Continuing with the example for series-parallel reduction

$$\left(\frac{1}{ans} + \frac{1}{86}\right)^{-1} = 78.03$$

which is realized in two steps as:

1 ÷ ans + 1 ÷ 86 [enter] → 0.013
∧ [(-)] 1 [sto→] X [enter] → 78.03

In a similar way, in two steps, we follow the steps for

$$\left(\frac{1}{300} + \frac{1}{256}\right)^{-1} = 138.13$$

The situation for variables ans and x at intermediate step is clearly shown in Fig. 5.5(c). Therefore,
+ x [enter] → 216.16 reduces the series connection.
This result is in parallel with the resistance of 120 Ω, so we find ans||120 = 77.16 Ω. The final result becomes then

R_{eq} = ans + 250 + 45 = 372.16 Ω.

The step by step process to this point is illustrated in Fig. 5.6.

5.1.3 Using Lists in series-parallel

We can use lists to calculate series and parallel equivalents. Defining

$$\mathbf{R} = \{R_1, R_2, \ldots, R_n\}$$

we proceed with formulas es follows:

For Series Connection: $R_{eq} = \text{sum}(\mathbf{R}), \quad G_{eq} = \dfrac{1}{\text{sum}(\mathbf{R})}$ (5.4a)

5.1. SERIES AND PARALLEL CONNECTIONS

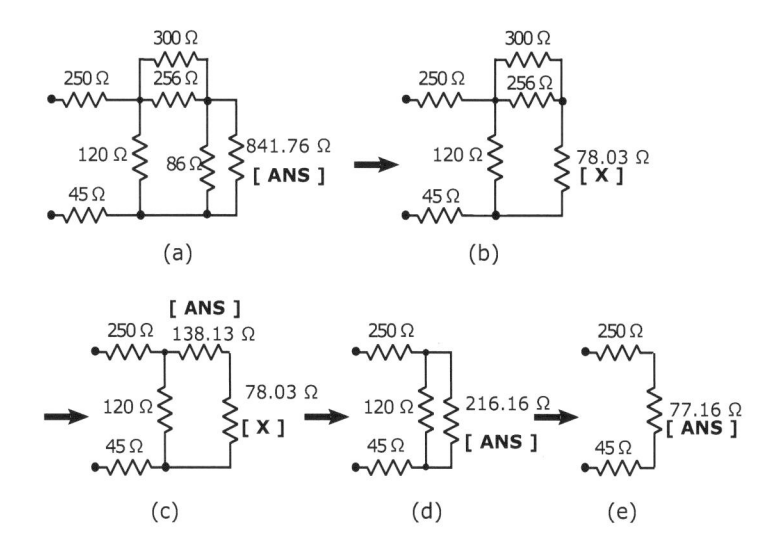

Figure 5.6: Example for series-parallel reduction

For Parallel Connection: $G_{eq} = \text{sum}(1/\mathbf{R})$, $\quad R_{eq} = \dfrac{1}{\text{sum}(1/\mathbf{R})}$ \hfill (5.4b)

Remember that the function sum() is in the statistics menu. It can also be typed directly.

The advantage of using lists is that it is less prone to mistakes and allows more complex calculations too when combined with memory or with stack. It may be or not faster, depending on particular situations. In general, though, it allows better control if you are using the list in a broader context. Let us start with the basic examples previously presented.

Example 5.3 *For Fig. 5.2(a), we define*

{526, 834} [sto→] X

to find

1/sum(1/X) → 322.56

For case (b)

{1.2E3,5.6E3,850} [sto→] X

to find

1/sum(1/X) → 456.96

For case (c),

$\{1.2E3 + 735, 3.21E3, 632+1.28E3\}$ `sto→` Y

to find

1/sum(1/Y) + 357 → 1.097E3

5.1.4 Function Parallel pl(z)

You could argue that lists are not always convenient, and you may be right. However, it is easy to define a function for calculation of the equivalent resistance of a parallel combination with lists as follows [2]:

$$\text{DEFINE } \text{pl(z)} = 1 \div \text{sum}(1 \div z) \quad (5.5)$$

where z is the list of resistances in parallel.

Remember, you can also use the store or assign (:=) method.

One advantage of functions is that you can use them directly as another value. We shall see the convenience of this function in the next example.

Example 5.4 *We want to find the equivalent resistance for the one port of Figure 5.7. All resistance values are in ohms (Ω), so the equivalent resistance will also be in ohms. Or it could be in kilohms, in which case the same units apply to the answer :)*

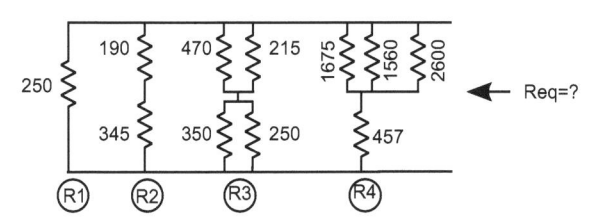

Figure 5.7: Example for Req calculation. Values are in Ω. Individual branches are denoted with circled labels.

The subcircuit consists of four branches in parallel, which are identified by circled R's; branches R2, R3 and R4 are themselves series and series-parallel sub circuits. Since

$$R2 = 345 + 190$$

we can use the expression directly. Now let us work the complete subcircuit, using auxiliary variables for easiness.

First create the individual branches separately and then create the list {250, R2, R3, R4}. *This method is easier to correct if an error is introduced.*

---
[2] If you have doubt on function definition, review Section § 2.7 on page 18

5.1. SERIES AND PARALLEL CONNECTIONS 61

I use the stack or auxiliary variables for R3 and R4. Execute the following steps. Answers are shown with three decimal figures.

Step 1 *Store R3 as Y:*

pl({350, 250}) + pl({470, 215}) sto→ Y enter → 293.352

Step 2 *Enter R4:*

457 + pl({1675, 1560, 2600}) enter → 1086.846

Step 3 *Create final list:*

{250, 345+190, Y, ans} sto→ X →
{250.000 535.000 293.352 1086.846}

Step 4 *Equivalent resistance:*

pl(ans) enter *or, equivalently,* 1/sum(1/ans) → 98.057

Let us take the next example using lists again to show the advantage of these graphical calculators. However, the important fact I want to bring to your attention is the convenience **to plan the strategy** before using the calculator, so you can make problem solving more effective. In this particular example, it illustrates a case where the reduction step is only part of the process. There is another part which consists in "going back" through the procedures to obtain the required answers.

This next example, on the other hand, provides a very nice application of lists. Try the example without using lists to see the difference! I have also used the previously defined function pl(z), but you may substitute it by the respective list operation if you prefer.

Example 5.5 *Find the voltage and current at each resistor in Fig. 5.8. Configure the calculator mode in* ENG. *That is, results are expressed engineering format display mode.*

Solution *Let us redraw the circuit with currents and voltages as shown in Fig. 5.9 (a). From here we proceed to set up our strategy.*

Step 1. *Since the 1.5, 2.3, and 3.8 kilo ohm resistors are in parallel, and thus with the same voltage V_1, we will be able to find the respective currents with Ohm's Law in one step using a list as*

$$\{I_1, I_2, I_3\} = V_1/\{1.5E3, 2.3E3, 3.8E3\} \tag{5.6}$$

Therefore, we first create the list of these resistances in parallel.

Step 2. *Use the list to obtain an equivalent resistor* Rx *using eq. (5.4b) and reduce the circuit to that of Fig 5.9 (b).*

Step 3. *Looking at this figure, we see that V_1 and V_4 will be found using (5.7) once we have I_4, while this one is obtained after knowing V_5, as shown in (5.8):*

CHAPTER 5. TRANSFORMATIONS AND REDUCTIONS

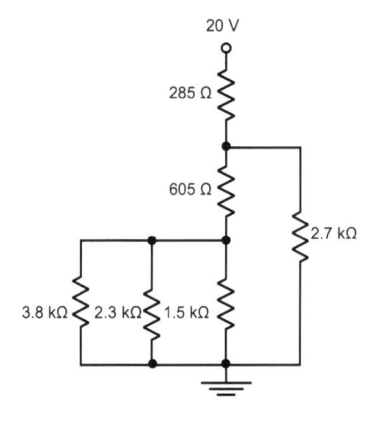

Figure 5.8: Another example with series-parallel reduction

$$\{V_1, V_4\} = I_4 \times \{Rx, 605\} \tag{5.7}$$

$$\{I_4, I_5\} = V_5 \div \{Rx + 605, 2700\} \tag{5.8}$$

Therefore, we will need the lists shown in these equations.

Step 4. With the last list, we obtain the equivalent resistor Ry arriving at Fig. 5.9 (c). We can see here that

$$\{V_6, V_5\} = I_6 \times \{285, Ry\} \tag{5.9}$$

so we create another list with these resistances. Now we find I_6 with

$$I_6 = 20 \div (285 + Ry) \tag{5.10}$$

We are now ready to proceed. In parenthesis, on the right of the step, I have included the meaning of the result for this example. Remember that if you do not want to use the function pl you may simple apply the corresponding formula.

5.1. SERIES AND PARALLEL CONNECTIONS

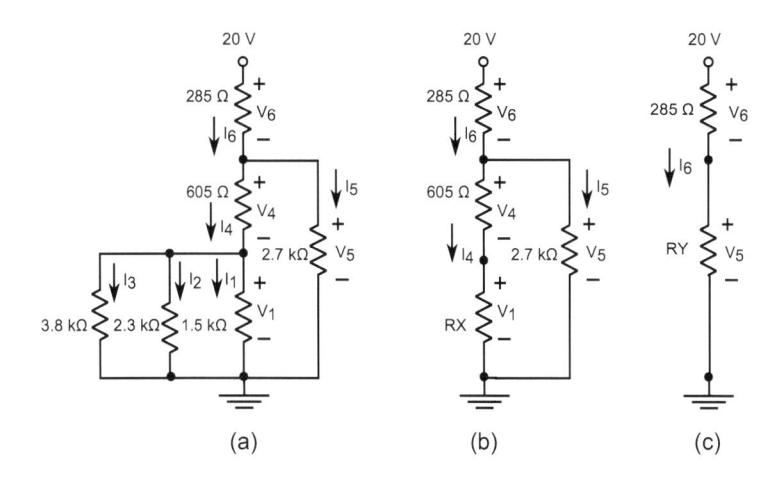

Figure 5.9: Another example with series-parallel reduction

1. **Create list L1 for Step 1:**
   {1.5 [EE] 3, 2.3 [EE] 3, 3.8 [EE] 3} [sto→] L1 [enter]

2. **Calculate Rx for step 2 and store as X:**
   pl(L1) [sto→]   X [enter] → 732.81E0   (Rx = 732.81 Ω)

3. **Create list L2 for (5.8):**
   {X + 605, 2.7 [EE] 3} [sto→]   L2 [enter]

4. **Calculate Ry for step 4:**
   pl(L2) [sto→] Y [enter] → 894.57E0   (Ry = 894.57 Ω).

5. **Calculate I_6 with (5.10):**
   20 ÷ (285 + Y) [enter] → 16.955E-3   (I_6 = 16.96 mA)

6. **Calculate V_6 and V_5 with (5.9):**
   × {285, Y} [enter] → {4.8323E0   15.168E0}
   (V_6 = 4.83 V and V_5= 15.17 V)

7. **Calculate I_4 and I_5 with with (5.8):**
   ans[2] ÷ L2 [enter] → {11.338E-3   5.6177E-3}
   (I_4 = 11.34 mA and I_5= 5.62 mA)

8. **Calculate V_1, V_4 with (5.7):**
   ans[1] × {X, 605} [enter] → {8.3084E0   6.8593E0}
   (V_1 = 8.31 V and V_4= 6.86 V)

9. **Calculate I_1, I_2, and I_3 with (5.6):**
   ans[1] ÷ L1 [enter] → {5.5389E-3   3.6123E-3   2.1864E-3 }
   (I_1 = 5.54 mA, I_2 = 3.61mA, and I_5= 2.19 mA)

An example with Homogeneity property

Let us take advantage of this example to see a nice application of the homogeneity principle.

Example 5.6 *For the circuit in Example 5.5 on page 61, find the voltages and currents in the circuit if the 20 V source is substituted by*
A) a voltage source of 5.6 V.
B) a voltage source of $15e^{-2t}$ V.
C) A voltage source V_s.

Solution:
The values of the currents and voltages for the original circuit are shown in the first columns of the two tables below. Let us obtain each table in one step. First, we define the column vectors using semicolon (;), using two decimals and currents scaled to mA units:

$$\mathbf{i} := [5.54;\ 3.61;\ 2.19;\ 11.34;\ 5.62;\ 16.96]$$

and

$$\mathbf{v} := [\ 8.31;\ 6.86;\ 15.17;\ 4.83]$$

You may perfectly use the template key [key] *or the editor.*

Now, the scaling factor for each of the cases A, B and C above are, respectively, 5.6/20, 15/20, and 1/20. Including the original case, let us create the row vector

$$\mathbf{f} := [1,\ 5.6/20,\ 15/20,\ 1/20]$$

The matrix multiplications $\mathbf{I} \times \mathbf{F}$ and $\mathbf{V} \times \mathbf{F}$ will produce the tables.
We can work the problem using lists too. I leave the exercise to the reader

Magnitude	Original 20 V	Case 5.6 V Factor $\frac{5.6}{20}$	Case $15e^{-2t}$ V Factor $\frac{15}{20}e^{-2t}$	Symbolic V_s Factor $\frac{1}{20}V_s$
I_1 (mA)	5.54	1.55	4.16 e^{-2t}	0.28 V_s
I_2 (mA)	3.61	1.01	2.71 e^{-2t}	0.18 V_s
I_3 (mA)	2.19	0.61	1.64 e^{-2t}	0.11 V_s
I_4 (mA)	11.34	3.18	8.51e^{-2t}	0.57 V_s
I_5 (mA)	5.62	1.57	4.22 e^{-2t}	0.28 V_s
I_6 (mA)	16.92	4.74	12.69 e^{-2t}	0.85 V_s

5.2. DELTA-WYE, WYE-DELTA TRANSFORMATION

Magnitude	Original 20 V	Case 5.6 V Factor $\frac{5.6}{20}$	Case $15e^{-2t}$ V Factor $\frac{15}{20}e^{-2t}$	Symbolic V_s Factor $\frac{1}{20}V_s$
V_1 (V)	8.31	2.32	$6.23e^{-2t}$	$0.42V_s$
V_4 (V)	6.86	1.92	$5.15e^{-2t}$	$0.34V_s$
V_5 (V)	15.17	4.25	$11.38e^{-2t}$	$0.76V_s$
V_6 (V)	4.83	1.35	$3.62e^{-2t}$	$0.24V_s$

5.2 Delta-Wye, Wye-Delta transformation

Let us go to the delta-wye (Δ to Y) and wye-delta (Y to Δ) transformations, which do not reduce the number of elements, but change the circuit topology. The delta (Δ) and wye (Y) subcircuits, shown in Fig. 5.10 a-b, are two useful topologies, and the transformations are specially important when working with complex impedances.

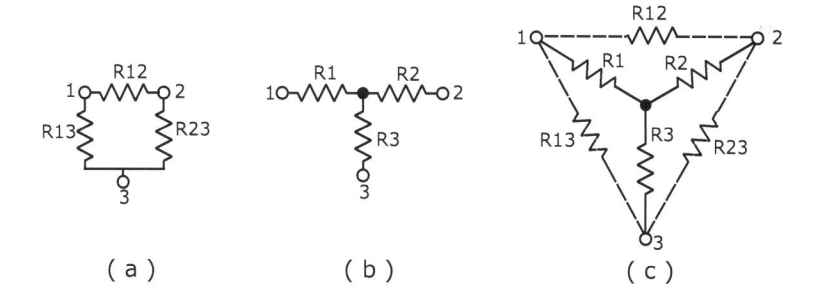

Figure 5.10: (a) Delta or Pi configuration; (b) Wye or T configuration; (c) Mnemotecnic reference

The two subcircuits are equivalent when voltages V_{12}, V_{13}, and V_{23}, and currents entering into terminals 1, 2 and 3 from outside are identical. This means that from an external point of view, one can substitute one subcircuit by the other in mathematical procedures. The equivalences are also used in circuits to take advantage physically, like for example realizing high valued resistances in integrated circuits, using lower valued ones.

The equivalence of both subnetworks is satisfied when the following transformations apply.

Delta-to-Wye Transformations:

$$R_1 = \frac{R_{12} R_{13}}{R_{12} + R_{13} + R_{23}} \quad (5.11a)$$

$$R_2 = \frac{R_{12} R_{23}}{R_{12} + R_{13} + R_{23}} \quad (5.11b)$$

$$R_3 = \frac{R_{13}\,R_{23}}{R_{12} + R_{13} + R_{23}} \qquad (5.11c)$$

Wye-to-Delta Transformations:

$$R_{12} = \frac{R_1\,R_2 + R_1\,R_3 + R_2\,R_3}{R_3} \qquad (5.12a)$$

$$R_{13} = \frac{R_1\,R_2 + R_1\,R_3 + R_2\,R_3}{R_2} \qquad (5.12b)$$

$$R_{23} = \frac{R_1\,R_2 + R_1\,R_3 + R_2\,R_3}{R_1} \qquad (5.12c)$$

These formulas were use as examples to program in section 2.8 on page 20. I now discuss them under other perspectives.

Fig. 5.10c provides a reference view for the formulas. In the first set, notice that the numerator is the product of the two resistances of the Delta structure that are connected to the same terminal of the Y-structure resistance. In the second set, the denominator is the non connected resistance of the Y-structure to the respective Delta-structure resistance. Most textbooks use Ra, Rb, and Rc for the Delta resistances. I prefer double subscripts because the formulas are easier to remember. In the first set of equations, the common subscript in the resistances of the numerator gives the subscript for the result. In the second set, the subscript of the Y resistance in the denominator is the one not present in the subscripts of the Delta resistance.

5.2.1 Working formulas directly

You can work formulas (5.11) and (5.12). Use memory/stack for easiness, combine with lists for less steps.

Example 5.7 Delta to Wye: *Take a Delta subcircuit with resistances R_{12} = 651 Ω, R_{13} = 1200 Ω, R_{23} = 760 Ω. We proceed to calculate with formulas (5.11)*

Step 1- Denominator to stack (ANS): 651 + 1200 + 760 → 2611

Step2- Calculate Wye resistances:
　　{651*1200, 651*760, 760*1200}/ans → {299.2　　189.49　　349.29}
　　Interpret result as {R1, R2, R3}

Wye to Delta: *Take now a Y subcircuit with R_1 = 600 Ω, R_2 = 313 Ω, R_3 = 728 Ω. To calculate formulas (5.12):*

Step 1- Store Numerator:
　　600 * 313 + 600 * 728 + 313*728 → 8.5246E5

5.2. DELTA-WYE, WYE-DELTA TRANSFORMATION

Step2- Calculate Delta resistances:
    ans/{728,313,600} → {1171 2723.5 1420.8}
    Interpret result as {R12, R13, R23}

5.2.2 Using functions

Let us first create functions for transformations using a list as parameter. First the Delta-to-Wye:

$$\text{DEFINE pi2y(z)} = \{z[1]*z[2], z[1]*z[3], z[2]*z[3]\} \div \text{sum(z)} \quad (5.13)$$

where z is the list of Δ-resistances $\{R_{12}, R_{13}, R_{23}\}$. The output of the function is the Wye-resistance list $\{R_1, R_2, R_3\}$.

Similarly, create the function for Wye-Delta transformation:

$$\text{DEFINE y2pi(z)} = \frac{z[1]*z[2] + z[1]*z[3] + z[2]*z[3]}{\{z[3], z[2], z[1]\}} \quad (5.14)$$

where z is the list of Wye-resistances $\{R_1, R_2, R_3\}$. The output of the function is the Delta-resistance list $\{R_{12}, R_{13}, R_{23}\}$.

Let us look at the previous example.

Example 5.8 *For example 5.7:*

Δ *to* Y: pi2y({651,1200,760}) → {299.2 189.49 349.29}

Y *to* Δ: y2pi({600,313,728}) → {1171 2723.5 1420.8}

One advantage of using lists is the economy in keystrokes when you have defined the list before calling the function. This is convenient, for example, when the original resistances are in fact equivalent resistances of another connection, as illustrated in the following example.

Example 5.9 *Find the Y equivalent for the delta sub circuit in Fig.5.11(a). The delta resistances are in fact equivalents of series-parallel subcircuits.*

Solution: *For this example, each of the resistances to be used in the formulas is in fact the equivalent resistance of a subcircuit. Specifically, using as reference Fig. 5.10, it is seen that*
    - R12 *is* 4.20 kΩ *in series with the parallel connection of* 1 kΩ||1.6 kΩ,
    - R13 *is the parallel connection* 2.5 kΩ|| (1.3 kΩ+ 13 kΩ), *and*
    - R23 *is the series connection of* 5.61 kΩ *and* 1.4 kΩ.

68 CHAPTER 5. TRANSFORMATIONS AND REDUCTIONS

(a)

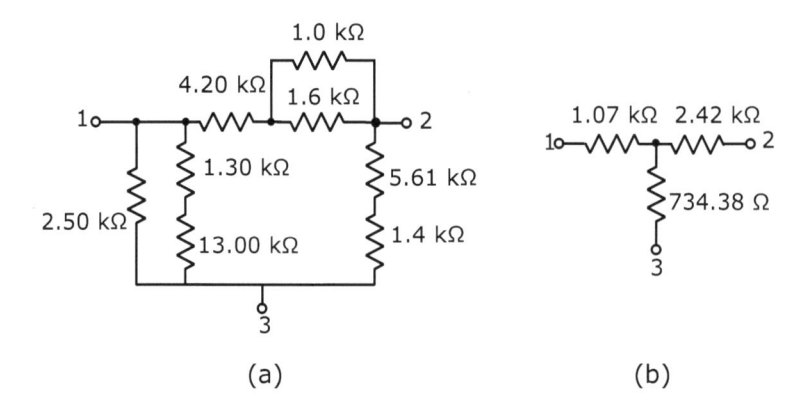

(b)

Figure 5.11: (a) Delta or Pi configuration to be transformed; (b) Equivalent Y configuration

Having identified these equivalencies, we introduce the equivalent resistances in the first step of transformation Delta-to-Wye using lists:

```
{(4.2E3 + 1/( 1/1E3 + 1/1.6E3 )), 1/(1 / 2.5E3 + 1/(1.3E3+13E3)),
(5.61E3+1.4E3)}  sto→  L6  →  {4.81E3, 2.13E3, 7.01E3}
```

The Y-circuit elements are then

```
pi2y(L6)  →  {1.07E3, 2.42E3, 734.38 }
```

The result should be interpreted as {R1, R2, R3 }, the configuration being that of Fig. 5.11(b).

If you prefer not to use lists in your parameter definition, you can always name your resistances explicitly name as shown next:

Our delta to wye function (or pi to T) is defined by

$$\text{pi2t(r12,r13,r23)} := \frac{\{r12*r13, r12*r23, r13*r23\}}{r12 + r13 + r23} \quad (5.15)$$

The output from this function should be interpreted as the list { R1, R2, R3} in the Wye configuration. Now, for the wye to delta (or T to pi) we can use

$$\text{DEFINE t2pi(r1,r2,r3)} = \frac{r1*r2+r1*r3+r2*r3}{\{r3,r2,r1\}} \quad (5.16)$$

The output from this function should be interpreted as the list { R12, R13, R23} of the delta configuration.

5.2. DELTA-WYE, WYE-DELTA TRANSFORMATION

The following example uses this notation and the `pl(z)` function.

Example 5.10 *Find the equivalent resistance for the subcircuit in Fig. 5.12. For display, use engineering mode with four digits.*

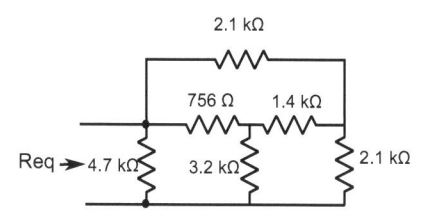

Figure 5.12: Example using Wye-Delta transformation

We can identify a wye configuration with the resistances of 756 Ω, 1.4 kΩ and 3.2 kΩ. To facilitate the identification of the terminals, let us introduce labels (1), (2) and (3) as shown in Fig. 5.13(a). Observe that after reducing the series connection of the parallel combinations, we still have R13 there to be used. Thus, we will store this value in the auxiliary variable X. The steps for solution are as follows:

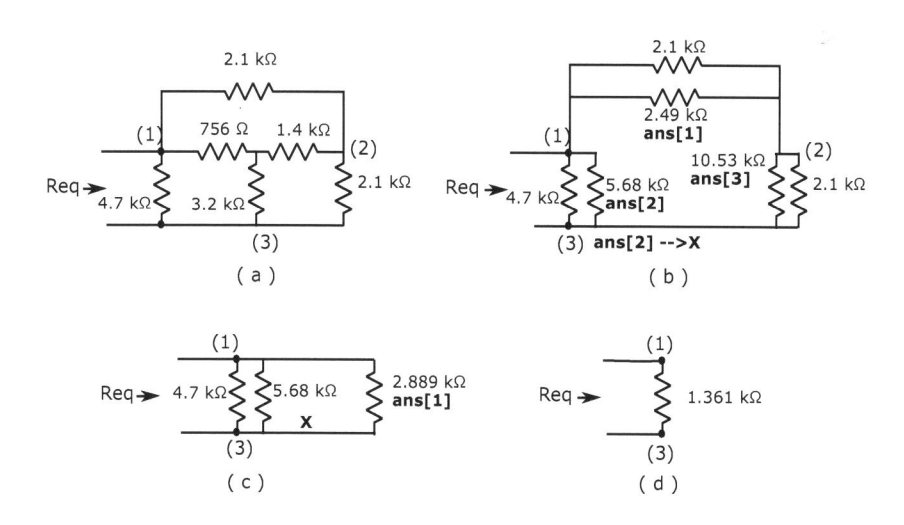

Figure 5.13: Steps in solving the example.

STEPS:

Step 1: Apply function t2pi:
t2pi(756,1.4E3, 3.2E3) → {2.487E3, 6.684E3, 10.53E3}
Result: Fig. 5.13(b)

Step 2: Use parallel function pl:
x:=ans[2]: pl({2.1E3, ans[1]})+pl({2.1E3, ans[3]}) → 2.889E3
Result: Fig. 5.13(c)

Step 3: Apply function pl again:
pl({4.7E3, x, ans} → 1.361E3
Result: Fig. 5.13(c); Final

5.3 Voltage and Current Dividers

Other configurations of great importance are the voltage and current dividers. Fig. 5.14(a) shows the usual two-resistor divider presented in most textbooks.

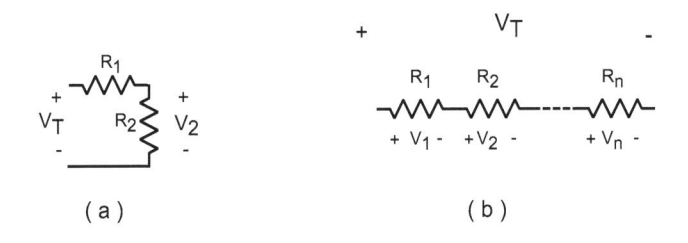

Figure 5.14: (a) Two-resistor voltage divider; (b) n-resistor voltage divider

The voltage at R_2 is given by

$$V_2 = \frac{R_2 V_T}{R_1 + R_2} \tag{5.17}$$

In the general n-resistor case shown in Fig. 5.14b, the voltage at any resistor R_j in the series string is given by

$$V_j = \frac{R_j V_T}{R_1 + R_2 + \ldots + R_n} \tag{5.18}$$

It is possible to immediately extend this formula with lists so we can calculate all the resistance voltages in one step. Defining the list $\mathbf{R} = \{R_1, R_2, \ldots, R_n\}$, we get

$$\{V_1, V_2, \ldots, V_n\} = V_T \times \mathbf{R} \div \operatorname{sum}(\mathbf{R}) \tag{5.19}$$

5.3. VOLTAGE AND CURRENT DIVIDERS

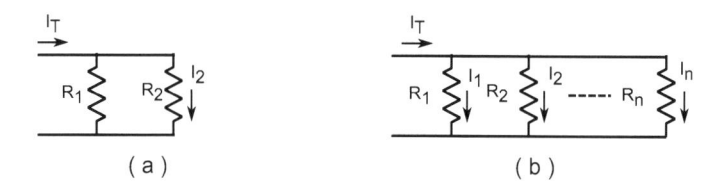

Figure 5.15: (a) Two-resistor current divider; (b) n-resistors current divider

The dual subcircuits of the previous voltage dividers are the current dividers shown in Fig. 5.15. For the two-resistor case we have

$$I_2 = \frac{G_2 I_T}{G_1 + G_2} \tag{5.20}$$

For two resistors, the formula is more commonly worked as

$$I_2 = \frac{R_1 I_T}{R_1 + R_2} \tag{5.21}$$

Notice that the numerator involves "the other" resistance. This formula is valid just for two resistors.

For the n-resistors divider we have

$$I_j = \frac{G_j I_T}{G_1 + G_2 + \ldots + G_n} \tag{5.22}$$

As in the case of voltage divider, it is also possible to express the current divider formulas using lists and in one step obtain all the resistor currents.

$$\{I_1, I_2, \ldots, I_n\} = I_T/\mathbf{R} \div \text{sum}(1/\mathbf{R}) \tag{5.23}$$

Let us start with simple examples.

Example 5.11 *For the voltage divider in 5.16(a), find V_1, V_2 and V_3. Also, find the currents for the current divider in (b). The calculator is in engineering format with two decimal figures for the results.*

Solutions *(a) For the voltage divider, we go through the following steps, using list variable R for faster procedure:*

R:= {1.6E3,2.1E3,1/(1/3.7E3+1/5.2E3)} → {1.60E3, 2.10E3, 2.16E3}

13 × R ÷ sum(R) [enter] → {3.55E0, 4.66E0, 4.79E0 }

The result shows the respective voltages V_1, V_2 and V_3 at the 1.6 kΩ, the 2.1 kΩ, and each of the resistances 3.7 kΩ and 5.2 kΩ in the parallel subcircuit.

72   CHAPTER 5. TRANSFORMATIONS AND REDUCTIONS

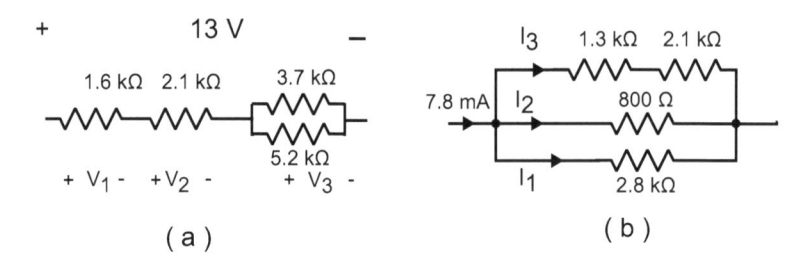

Figure 5.16: (a) Two-resistor voltage divider; (b) n-resistor voltage divider

(b) For the current divider, using variable R again we proceed as follows:

{2.8E3,800,1.3E3+ 2.1E3} sto→   R

7.8 E (-) 3 ÷ R ÷ sum(1/R) enter ⇒ {1.47E⁻3, 5.13E⁻3, 1.21E⁻3}

Thus, the current I_1 through the 2.8 kΩ is 1.47 mA, I_2 through the 800 Ω resistor 5.13 mA, and for the series branch $I_3 = 1.21$ mA.

Remark: *remember that by scaling units you may save keystrokes. In particular, using kilo ohms for resistances in this formula does not alter the division. In addition, use mA for current.*

The following example shows a situation in which the list denoting the series resistance must be chosen with a specific order because of the operations involved. It illustrates, once more, that using technology in problem solving requires also a good planning. In this example I use the Cumulative Sum List funcion, cumulativeSum() function for lists. This function is called with the sequence " menu 6 4 3".

Example 5.12 *Fig. 5.17a shows a string of resistances in series used often in applications, for example in flash type analog-to-digital converters. In this subcircuit, more than focused on individual resistance voltages, interest is on the node potentials V_j. These are expressed as*

$$V_j = \frac{R_1 + R_2 + \ldots + R_j}{\sum_{h=1}^{N} R_h} V_s$$

If we define $\mathbf{R} = \{R_1, R_2, \ldots, R_n\}$, *in that order,* **from ground to source**, *then we can find all potentials in one step as shown below. Observe in the formula that we include the source itself, which is the potential of node n. That is,* $V_n = V_s$. *The reader can modify the formula to exclude it*[3].

---

[3] See guidebook

5.3. VOLTAGE AND CURRENT DIVIDERS

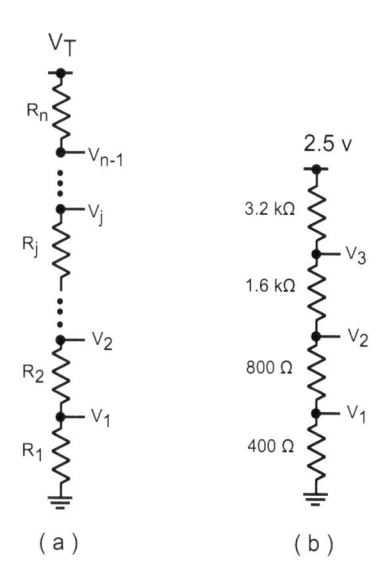

Figure 5.17: (a) A string voltage divider configuration; (b) Particular example.

$$\{V_1, V_2, \ldots V_n (= V_s)\} = V_s \times \text{cumulativeSum}(\mathbf{R}) \div \text{sum}(\mathbf{R})$$

Taking the numerical example in Fig. 5.17b, and using variable X for key-stroke savings, we have:

{400,800,1600,3200} `sto→` R
2.5× cumulativeSum(R) ÷ sum(R) `enter` → {.17, .50, 1.17, 2.50}

Therefore, $V_1 = 0.17$ V, $V_2 = 0.50$ V, *and* $V_3 = 1.17$ V

5.3.1 Voltage dividers again: using conductances

Applications of two-resistor dividers are very large, so I thought that another look at this particular case is convenient to illustrate advantages of the calculator when the theory is applied appropriately. In this subsection, the use of conductances is stressed since this formula is better for many situations. In particular, when one of the resistances is in fact the equivalent of a parallel connection, such as in loaded dividers or passive adders.

Consider the case of two-resistor voltage divider using the formula (5.17). we can multiply numerator and denominator by $G_1 G_2$ to obtain the expression also in terms of conductances shown below.

$$V_o = \frac{R_2 V_T}{R_1 + R_2} = \frac{G_1 V_T}{G_1 + G_2} \qquad (5.24)$$

Observe that the formula can be expressed as

$$V_o = \frac{G_1}{G_{eq}} V_T$$

where G_{eq} is the equivalent conductance of the resistances in parallel.

Loaded Divider

The formula in terms of conductances allows a direct treatment in the case of a loaded divider in Fig. 5.18. Now, G_2 becomes an equivalent conductance, and we can extend (5.24) with the following two steps to calculate the output voltage using lists for any number of parallel resistors:

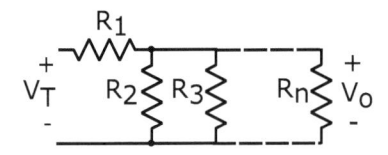

Figure 5.18: A loaded two-resistor type divider.

$$\begin{array}{l}\text{Step 1: Define } \mathbf{G}=1/\{R_1, R_2, R_3, \ldots, R_n\} \\ \text{Step 2: Calculate } V_o = V_T \times G[1] \div \text{sum}(\mathbf{G})\end{array} \qquad (5.25)$$

An alternative for step 1 is {1/R1, 1/R2, ...} [sto→] G, which has the advantage of dealing with infinite resistances (1/R=0).

Example 5.13 *Using three resistances in Fig. 5.18, with $R_1 = 350\ \Omega$, $R_2 = 950\ \Omega$, $R_3 = 1650\ \Omega$, and $V_T = 8$ V, then :*

1/{350,950,1650} [sto→] g [enter] → {2.857E-3   1.053E-3   606E-6}
8 [×] g[1] / sum(g) [enter] → 5.062E0

Thus, Vo = 5.062 V

Example 5.14 *In the previous example, assume that the load R_3 takes the different values 200 Ω, 300 Ω, 750 Ω, 2000 Ω, and 5000 Ω. Calculate the output voltage for each case, and also the unloaded case ($R_3 = \infty, G_3 = 0$).*

Solution: *Since in this case the unloaded divider should also be considered, let us introduce the list of loads directly as a list of conductances.*

{1/200,1/300,1/750,1/2000,1/5000,0} [sto→] GL [enter]

5.3. VOLTAGE AND CURRENT DIVIDERS

Then we apply (5.25) as

```
8* (1/350)/(1/350+1/950+GL) enter
```
→ {2.565   3.156   4.359   5.183   5.562   5.846}

The values in the final list are the respective output voltages for the different load conditions.

Let us work the example using dot operations of matrices, just for the sake of looking at something different. The procedure is a little more complicated, but the objective is to show another use of matrices.

Example 5.15 *To proceed, we follow the next steps:*

1. *Create a matrix* **G** *of all 1's, of order 3 x 6 (One column for each value of the load)*

2. *Multiply the first row by* $1/R_1$, *and the second row by* $1/R_2$.

3. *Substitute the third row by a row using conductances of* R_3

4. *Apply the formula of the second step in (5.25) using dot-operations:*
   V_T* `g[1]./sum(g)`

Now let's go to the steps. I do not show the results on display to save space, only the last one.

1. `subMat(3,6)` sto→ g: Fill 1,g enter

2. `g[1]/350` sto→ `g[1]`:   `g[2]/950` sto→ `g[2]` enter

3. `[1/200,1/300,1/750,1/2000,1/5000,0]` sto→ `g[3]` enter

4. `8* g[1]./sum(g)` enter
   → [2.565   3.156   4.359   5.183   5.562   5.846]

The values in the final vector are the respective output voltages for the different load conditions.

Remark: *The* `subMat()` *and the* `Fill` *functions are in the matrix Create submenu* menu *7 1. Observe that dot operations have been used.*

Passive adder with voltage divider: applying superposition

Owing to superposition, resistive circuits with many inputs are sometimes call adders, because they yield a weighted sum of inputs. An important case is the common passive adder of Fig. 5.19(a), which is basically an extension of the voltage divider.

We can apply superposition to get V_o here. When all inputs but V_j are zero, the configuration we are left with is that of Fig. 5.19(b), which is in fact

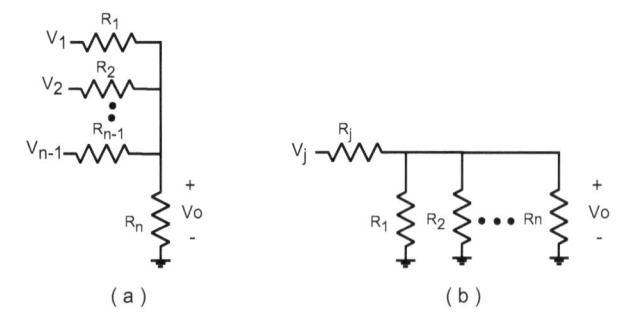

Figure 5.19: (a) A passive adder configuration; (b) When all inputs but V_j are zero (R_j not grounded)

a loaded divider similar to the one in Fig. 5.18. For this individual element we can write

$$V_{oj} = \frac{G_j}{G_{eq}} V_T \qquad (5.26)$$

where G_{eq} is the equivalent conductance of all resistances in parallel. If you prefer to use resistances in formulas, just for the record, this expression is sometimes expressed as

$$V_{oj} = \frac{R_{eq}}{R_j} V_T \qquad (5.27)$$

The use of resistances in notation does not bring up superposition in an easy format, useful for our purposes. However, with conductances we use (5.26) for each input, we apply superposition and get,

$$V_o = \frac{V_1 G_1 + V_2 G_2 + \ldots V_{n-1} G_{n-1}}{G_1 + G_2 + \ldots + G_{n-1} + G_n} \qquad (5.28)$$

In addition to simplicity in the expression, we can work the case $R_n = \infty$, where $G_n = 0$. Moreover, since this value appears only in the denominator, there will be cases in which you can ignore it.

On the other hand, looking at (5.28), you may notice that there are $n-1$ terms in the numerator but n in the denominator. To simplify formulas with lists, we need to comply with compatibility in dimension. Hence, the list for voltages includes a dummy 0 V source as V_n. You can think of this dummy source to be in series with resistance R_n. If $R_n = \infty$, you may not include this dummy voltage and neither the conductance G_n; that is, both lists will be of dimension $n-1$ for $R_n = \infty$.

Let us define the two lists of resistances and sources:

1. **R**=$\{R_1, R_2, R_3 \ldots R_n\}$

5.3. VOLTAGE AND CURRENT DIVIDERS

2. $\mathbf{V} = \{V_1, V_2, V_3 \ldots V_{n-1}, 0\}$

We can now calculate equation (5.28) in one or two steps. I'll mention it as a two step process to illustrate further applications.

1. Calculate list of contributions: $\mathbf{V}_o = \dfrac{\mathbf{V} \div \mathbf{R}}{\text{sum}(1/\mathbf{R})}$ (5.29a)

2. Sum the elements of the list: sum(Vo) (5.29b)

Notice that, discarding the last element, (5.29a) provides the individual contributions of the sources while (5.29b) applies superposition.

We are now in position to obtain different results easily:

A) To find the individual weights Use all sources as 1 Volt, so the individual weights are found as

$$V_o = 1 \div \mathbf{R} \div \text{sum}(1/\mathbf{R}) \quad (5.30)$$

Discard the last element of the list for finite R_n, .

This is in particular useful when symbolic sources are present.

B) To find individual contributions Define list $\mathbf{V}$, and use (5.29a) to find individual contributions, after discarding the last element in he list for finite R_n. Express the result using superposition. When sources are of the form $k_j f_j(t)$, use the coefficients in the definition and interpret the final result yourself by superposition

C) To Find the total output for numerical inputs That is, to calculate (5.28), you may either follow the two steps above (which have the advantage of providing individual contributions) or else do it in one step as

$$V_o = \text{sum}(\mathbf{V} \div \mathbf{R}) \div \text{sum}(1/\mathbf{R}) \quad (5.31)$$

Let us take an example to look at the different alternatives.

Example 5.16 *In Fig. 5.19(a), consider n=3, with R_1=210 Ω, R_2 = 150 Ω, and R_3 = 2 kΩ. and two inputs V_1 and V_2 connected to R_1 and R_2, respectively. For this adder,*

1. *Express the ouput as the weighted sum of inputs.*

2. *Find the individual contributions and the total output voltage Vo when the inputs are $V_1 = 2.1$ V, $V_2 = 1.8$ V.*

3. *Find Vo if the inputs are $V_1 = 2.1 \cos(250t)$ V, $V_2 = 1.8\ e^{-100t}$ V*

Solution: For all items, we need the resistance list. Let us store the list to R

{210, 150, 2E3 } [sto→] r

The solutions follow next:

1. Enter   1/r/sum(1/r) → {0.399   0.589   0.042}
   Discard the last element and interpret: $V_o = 0.399\,V_1 + 0.589\,V_2$.

2. Create list for the inputs V, {2.1, 1.8, 0 } [sto→] v

   For individual contributions: v/r/sum(1/r) → {0.838   1.006   0}.

   Discarding last item, the individual contributions for the 2.1 V and 1.8 V sources are, respectively, 0.838 V and 1.006 V

   For output: sum(ans) → 1.844.   Thus, $V_o = 1.844$ V.

3. From the individual contributions in the previous item, we arrive at

$$V_o = 0.838\cos(250t) + 1.006 e^{-100t} \quad V$$

Let us close this section with some comments about the passive adder. Many times students under appreciate "simple" circuits until they come to understand that the usefulness comes not from simplicity or complexity, but by a proper interpretation of (5.28). Let us write again this equation to look at it with a renovated approach.

Let us write the equation as

$$V_o = k_1\,V_1 + k_2\,V_2 + \ldots k_{n-1}\,V_{n-1} \tag{5.32}$$

where

$$k_j = \frac{G_j}{G_T} = \frac{R_T}{R_j} < 1; \quad \text{and} \sum_{j=1}^{n-1} k_j \leq 1$$

R_T is the equivalent resistance for all resistances in parallel. The equality sign applies when $R_n = \infty$.

Let us interpret this formula in several applications:

- **Average Value**: If $R_n = \infty$ and all resistances are equal, than Vo has the average value of all inputs.

5.3. VOLTAGE AND CURRENT DIVIDERS

- **DC-shift**: Again, with $R_n = \infty$, and with only two inputs, if one is time dependent, like $\cos(\omega t)$ and the other is a DC constant value source, then this source provides a controlled constant shift of the time varying function.

- Summing is precisely what is done in many "mixers" used in communication systems for different applications.

To correct for the attenuation factor, the adder may be followed by an amplifier, as shown in Fig. 5.29.

To illustrate a mixer application, let us look at an "amplitude modulator" example.

Example 5.17 *This example is based on [Budak87]. In Fig. 5.19(a), take n=2 resistances, with $R_1 = R_2 =$ and two inputs $V_1 = V_m \sin(\omega t)$ V and $V_2 = V_m \sin(\omega t + \Delta \omega t)$ V connected to R_1 and R_2, respectively. We assume here that the magnitude of $\Delta \omega$ is such that both inputs are nearly equal in frequency. For this adder,*

$$V_o = \frac{1}{2}\left(V_m \sin(\omega t) + V_m \sin(\omega t + \Delta \omega t)\right)$$

Since $\sin A + \sin B = 2 \cos(\frac{1}{2}[A - B]) \sin(\frac{1}{2}[A + B])$ *we have*

$$V_o = V_m \cos\left(\frac{\Delta \omega}{2} t\right) \sin\left(\omega t + \frac{\Delta \omega}{2} t\right)$$

This is a sine wave of frequency $(\omega t + \frac{\Delta \omega}{2})$ whose amplitude varies slowly at the rate $\Delta \omega/2$. We say that the amplitude is modulated. To see this effect, we can graph the figure on our calculator. Let us take $V_m = 3V$, our frequency $\omega = 600$ rad/s and $\Delta \Omega = 50$ rad/s. You may open the graph window hitting the home key and then B.

With menu *4 1 you open the Windows settings menu. Let us put the window settings at* xmin =0, xmax = $2\pi/25$, ymin = -4 ymax = 4 *for a graph with one cycle of the envelope* $3 \cos(25t)^4$.

With these settings we proceed to define our functions. With menu *4 1 you call the entry line. Type the function f1(x) below, and then the cursor key* ▼ *to go to the next function. When you have finished, you will see the graph in Fig. 5.20. I cleaned the screen taking out the labels. ;)*

```
f1(x) = 1.5 sin(250*x) + 1.5 sin(300*x)
f2(x) = 3 cos(25*x)
f3(x) =  (-)  f2(x)
```

---

[4]Since the frequency is 25 rad/s, then each period is $2\pi/25$ s. Hence, xmax has been set to cover one period, or cycle, of the function.

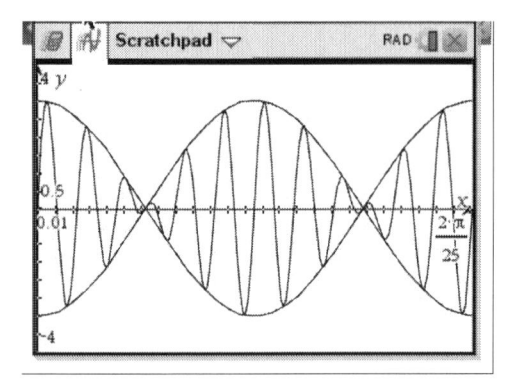

Figure 5.20: Passive adder output for the modulation example.

5.3.2 Thevenin equivalent for two-resistors divider.

Another useful transformation is that of the Thevenin's equivalent for a voltage divider (Fig. 5.21. The equations for the Thevenin voltage V_t and resistance R_t in this case are

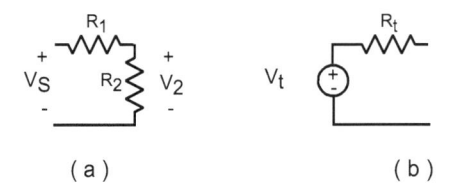

Figure 5.21: Thevenin equivalent circuit for a particular case

$$V_t = \frac{R_2 V_s}{R1 + R2} \qquad (5.33)$$

$$R_t = \frac{R_1 R_2}{R1 + R2} = R_1 \| R_2 \qquad (5.34)$$

We see that they are in fact the same as (5.24) and the formula for parallel resistance, respectively. We combine both in a function that yields $\{V_t, R_t\}$:

$$\text{Define th(V,R1,R2) = R2} \times \{V, R1\} \div (R1 + R2) \qquad (5.35)$$

Example 5.18 *For the ladder attenuator of Fig. 5.22, find the attenuation factor and the output resistance.*

By attenuation factor in a resistive circuit you mean the ration Vout/Vin, *which will be less than 1. If you make* Vin = 1 V, *then the output voltage*

5.3. VOLTAGE AND CURRENT DIVIDERS

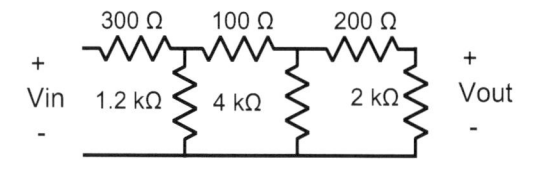

Figure 5.22: Ladder attenuator example.

will be numerically equal to the desired factor. This means that from the algorithmic point of view we are actually looking for is the Thevenin equivalent at the output for a voltage input of 1 V. Then V_t becomes the attenuation factor and R_t the output resistance.

The steps are shown in the table below with three decimal places, and the successive reductions in Fig. 5.23. These steps are mentioned in the right column.

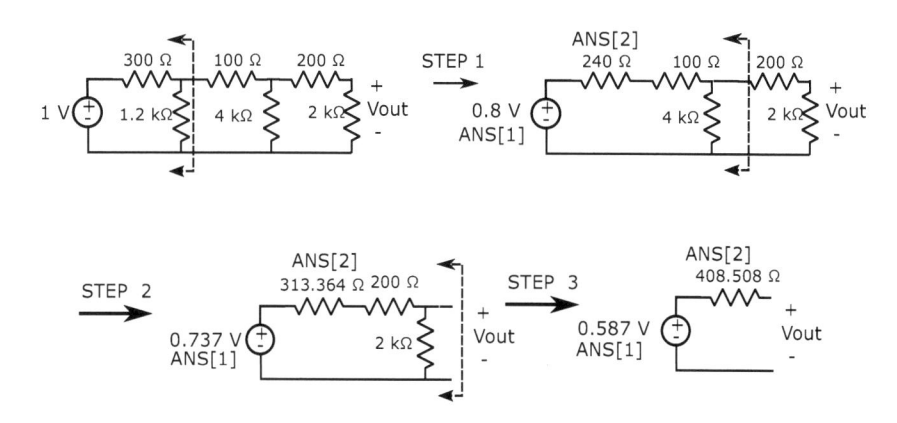

Figure 5.23: Reduction steps for the ladder attenuator example. (Stack position shown for intermediate and final values)

Command Line	Result in Stack	Reduction step
th(1,300,1200) enter	{0.8, 240}	Step 1
th(ans[1],ans[2]+100,4000) enter	{0.737, 313.364}	Step 2
th(ans[1],ans[2]+200,2000) enter	{0.587, 408.508}	Step 3

The attenuation factor is thus 0.587 and the output resistance 408.51 Ω.

5.4 Source transformations

Source transformations, illustrated in Fig. 5.24 are very useful and common in applications. For these transformations, we apply

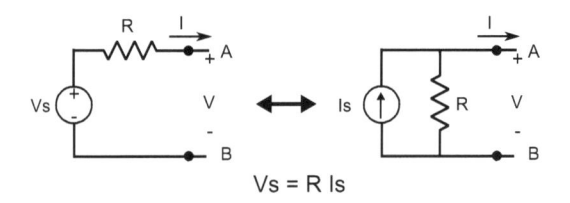

Figure 5.24: Source transformation theorem

$$I_S = \frac{V_S}{R} \tag{5.36}$$

for voltage-to-current source transformation, and

$$V_S = R I_S \tag{5.37}$$

for current-to-voltage source transformation. Notice the great similarity with Ohm's law.

When convenient, we can realize several transformations at once using lists, as illustrated with the following example.

Example 5.19 *Consider the circuit in Fig. 5.25(a), where we want to find the current I_x. This example uses auxiliary list variables for storage of intermediate results, illustrated in the different insets of Fig. 5.25. To solve the problem we follow the next steps.*

1. **Convert $5V$ - 8Ω (left) and $8V$ - $14\ \Omega$ (right):**   *[Fig. 5.25(b)]*
   $\{5,8\}$ / $\{8,14\}$ enter $\rightarrow \{.625 \quad .571\}$

2. **Reduce parallel sources and resistances:**   *[Fig. 5.25(c)]*
   ans+$\{0.5,$ (-) $0.3\}$ sto$\rightarrow$ L1 enter $\rightarrow \{1.125 \quad .271\}$

   $\{1/(1/8+1/20)$, $1/(1/7+1/14)\}$ sto$\rightarrow$ L2 enter $\rightarrow$
   $\{5.714 \quad 4.667\}$

3. **Convert sources and combine resistances:**   *[Fig. 5.25(d)]*
   ans $\times$ L1 sto$\rightarrow$ L1 enter $\rightarrow \{6.429 \quad 1.267\}$
   L2 + $\{6,10\}$ sto$\rightarrow$ L2 enter $\rightarrow \{11.714 \quad 14.667\}$

5.4. SOURCE TRANSFORMATIONS

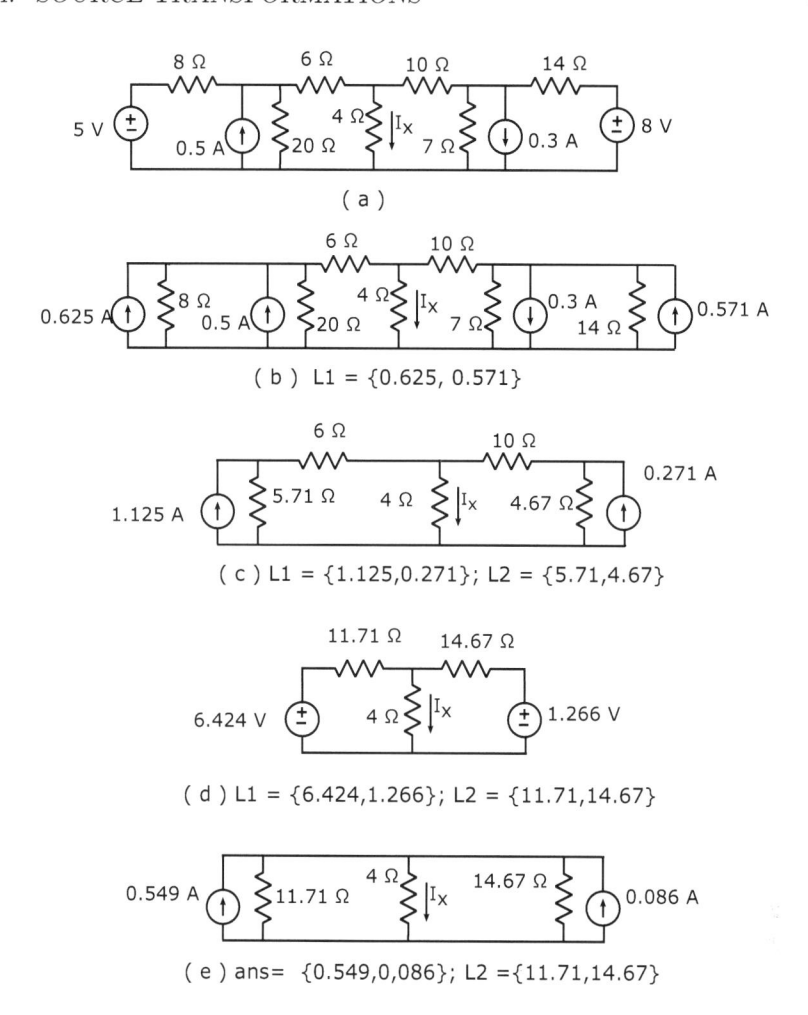

Figure 5.25: Circuit for example 5.19 on source transformation

4. Convert again: *[Fig. 5.25(e)]*

L1 $\div$ L2 enter → {.549 .086}

Final Step: Find current by current division

sum(ans)/4 / (1/4 + sum(1/L2)) enter → .393

We see once again in these steps the convenience of lists and of user defined functions to simplify procedures and speed up problem solving. Observe also that using Ohm's law with conductances instead of resistances simplified the number of steps.

5.4.1 Shifting theorems, and extended source transformations

Two theorems rarely mentioned in basic textbooks are the shifting theorems illustrated in Fig. 5.26. Here, all currents and voltages inside subcircuit, as well as the potential at the terminal nodes, and the currents entering the subcircuit from these nodes remain unchanged. Notice that the theorems can be applied in both directions, either to split into several sources, or to reunite several onto one. Moreover, they are applicable to dependent sources.

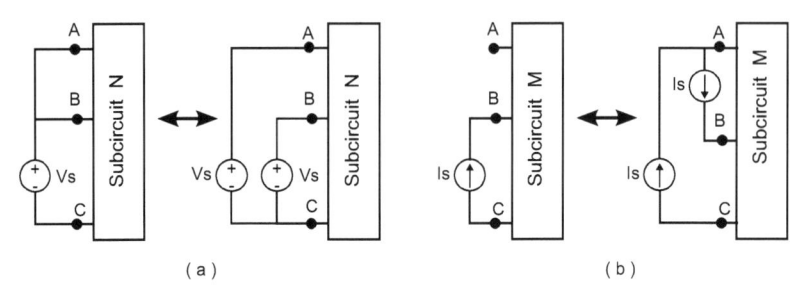

Figure 5.26: Source shifting theorems. (a) Voltage shifting theorem (b) Current shifting theorem

I work now only the voltage shifting, and leave to the reader working with the dual theorem. Fig. 5.27 on the facing page shows two transformations applying the voltage shifting theorem.

In the case of (a), multiple voltage-to-current transformations are carried out at once using lists. In the second case (b), several Thevenin transformations are obtained in few steps, also using lists. The procedures are indicated in the figure themselves. The relations shown in Fig. 5.27 can be of course extended to any number of sources. Let us illustrate with one example.

Example 5.20 *Consider the bridged T shown in Fig. 5.28(a), where we want to find current I. The application of the transformation shown in Fig. 5.27(b) results in Fig. 5.28(b). Calculations are as summarized in the following table:*

Entry	On stack	Comment
{300, 200} sto→ R1	{300. 200.}	
{1200, 2000} sto→ R2	{1200. 2000.}	
1 × R2 ÷ (R1 + R2) sto→ V	{.8 .9090909}	{Va, Vb}
R1 × R2 ÷ (R1 + R2) sto→ Rt	{240 181.8182}	{Rta, Rtb}
(V[2]-V[1]) ÷ (100 + sum(Rt))	.0002091	Current

Therefore, I = 209.1 μA. Observe that reading would have been easier using ENG settings.

5.4. SOURCE TRANSFORMATIONS

{ Ia, Ib } = Vs / { RA, RB }

(a)

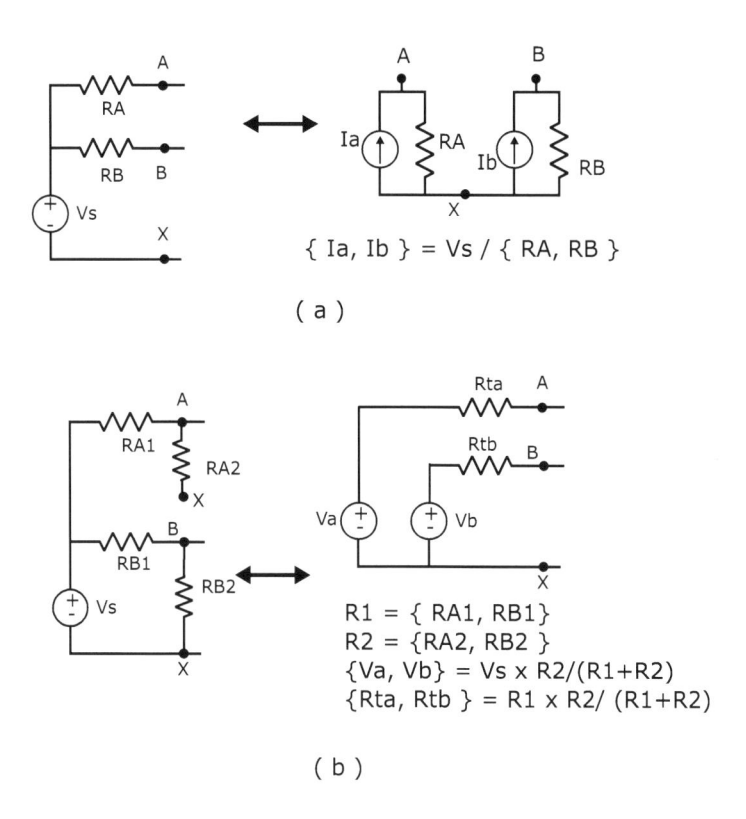

R1 = { RA1, RB1}
R2 = {RA2, RB2 }
{Va, Vb} = Vs × R2/(R1+R2)
{Rta, Rtb } = R1 × R2/ (R1+R2)

(b)

Figure 5.27: Applying source shifting theorems (a) voltage shifting to current source transformations, (b) voltage shifting for several Thevenin equivalents computation.

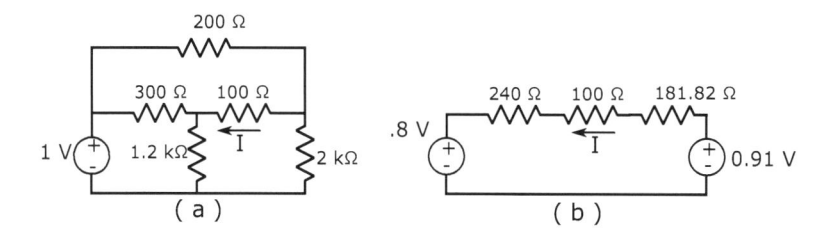

Figure 5.28: Thevenin reduction for a bridged-T.

Again, use of lists and application of theorems need practice, practice, and practice. Use different exercises from your textbooks.

5.5 Some operational amplifier structures

Blocks with operational amplifiers are extremely useful and have multiple applications. Basic operational amplifier structures include the common non-inverting and inverting amplifiers, but there are of course many many other structures to explore!

The next example uses the structures in Fig. 5.29. With lists, we can apply superposition and also do fast computations with calculator. Observe that these examples can be reduced to the the basic amplifier structures as well as the subtractor.

In the example, I do not show the calculator steps, but only the concepts. Lists are represented with bold fonts, while scalars with non bold ones.

Example 5.21 *Fig. 5.29 shows three common operational amplifier structures: the inverting adder, the non-inverting adder, and the adder-subtracter configurations. For simplicity, let us work with three inputs for the first two cases, and three adding voltages with three voltages to be subtracted in the third case. Equations and procedures, however, are valid for any number of cases, including the special one input amplifiers.*

Let us work each case.

(A) For the inverting adder, the mathematical formula is :

$$V_{out1} = -\left(\frac{R_f V_1}{R_1} + \frac{R_f V_2}{R_2} + \frac{R_f V_3}{R_3} +\right) = R_f \sum_{j=1}^{3} \frac{V_i}{R_i} \quad (5.38)$$

To work in the calculator, define the lists $\mathbf{R} = \{R_1, R_2, R_3\}$ *and* $\mathbf{V} = \{V_1, V_2, V_3\}$. *Then the individual terms of (5.38) are found with*

$$\mathbf{V_{out}} = -R_f \times (\mathbf{V}/\mathbf{R}) \quad (5.39)$$

This step becomes handy for superposition, symbolic inputs and so on. The numerical output written as

$$v_{out} = \mathrm{sum}(\mathbf{V_{out}}) \quad (5.40)$$

(B) The non-inverting adder is basically a passive adder with a non-inverting amplifier. Therefore, we use (5.28) multyplying it by the gain:

$$V_{out} = \frac{(V_1 G_1 + V_2 G_2 + V_3 G_3)}{G_1 + G2 + G_3 + G_g}(1 + R_f/R) \quad (5.41)$$

Define $\mathbf{R} = \{R_1, R_2, R_3, R_g\}$ *and* $\mathbf{V} = \{V_1, V_2, V_3, 0\}$. *Then, to obtain the individual terms for this formula apply*

$$v_{out} = (1 + R_f/R) \times (\mathbf{V}/\mathbf{R})/\mathrm{sum}(1/\mathbf{R}) \quad (5.42)$$

5.5. SOME OPERATIONAL AMPLIFIER STRUCTURES 87

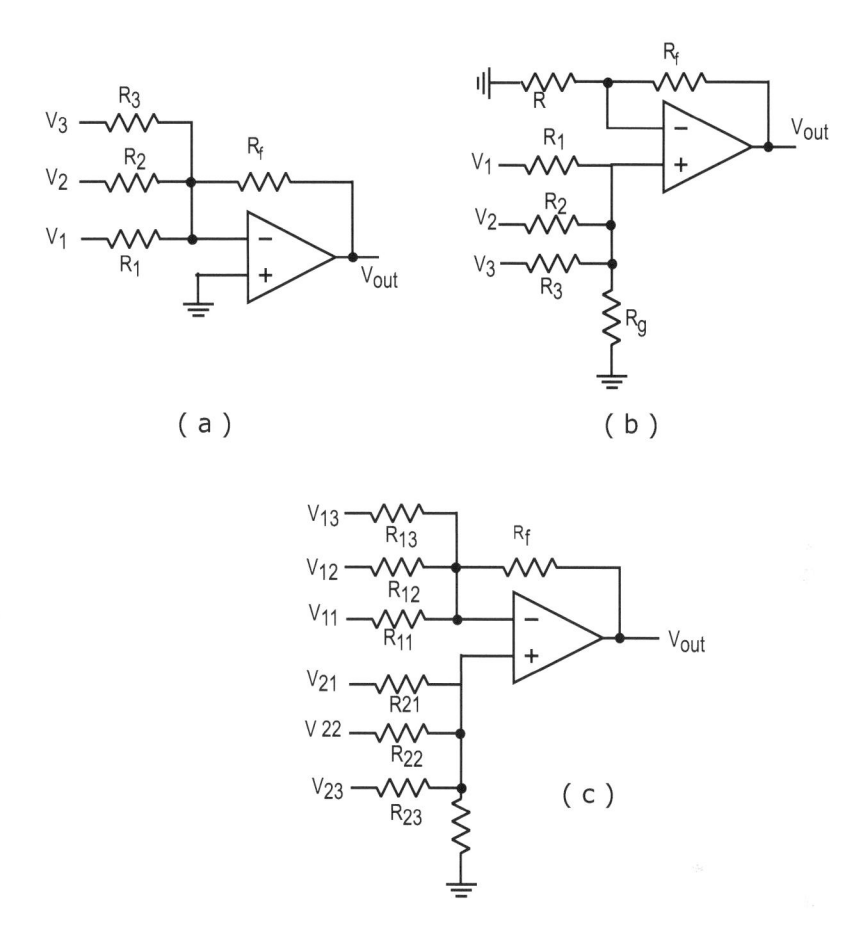

Figure 5.29: Three OA structures: (a) inverting adder; (b) non-inverting adder; (c) adder-subtracter

and sum($\mathbf{V_{out}}$) to get the numerical output voltage

(C) Finally, for the adder-subtracter, we combine (5.38) and (5.41), where R is the parallel combination of $R_{11}, R_{12}, \ldots$. The sequence in the calculator, using variables (X, Y, R, A, B) for simplicity, may be as indicated below:

1. $\{R_{11}, R_{12}, R_{13}\}$ [sto→] Rn

2. 1÷sum(1/Rn) [sto→] R

3. $\{V_{11}, V_{12}, V_{13}\}$ [sto→] Vn

4. [(-)] R_f× (Vn/Rn) [sto→] A

h!t

```
1   Define addsub(rf,gn,vn,gp,vp)=
2   Func
3   Local geq,nc,pc
4   geq:=sum(gn)
5   If vn={ } Then
6   nc:={ }
7   ElseIf geq=0 Then
8   nc:={ }
9   Else
10  nc:=rf*gn*vn
11  EndIf
12  If vp={ } Then
13  pc:=
14  ElseIf gp={ } Then
15  pc:=(1+rf*geq)*vp
16  Else
17  pc:=(((1+rf*geq)*vp*gp)/(sum(gp)))
18  EndIf
19  Disp "Vout Components"
20  augment(pc,nc)
21  EndFunc
```

Figure 5.30: An example of a function for OA adder-subtracter.

5. $\{R_{21}, R_{22}, R_{23}, R_g\}$ sto→ Rp

6. $\{V_{21}, V_{22}, V_{23}, 0\}$ sto→ Vp

7. (1 + R$_f$/R)× (Vp/Rp)/sum(1/Rp) sto→ B

8. *Voltage output components:* augment(B,A) sto→ vout

9. *Numeric output voltage* sum(vout)

The reader is encouraged to try numeric values for data.

5.5.1 A function for the OA structures

For personal use I wrote the function in Fig. 5.30. Several comments apply

1. Parameters gn, vn, gp, vp are lists. Parameter rif can be 0. Lists vn, gp, vp can be void ({ }), but not gn.

2. Except for resistance rif, all others must be entered as conductances. Infinite resistances are entered as conductance 0

3. The result is a list of components of Vout, NOT the numeric value of the output voltage. Use `sum(ans)` right after the function if you want a numeric value

4. The result may include a 0 valued element which the user must ignore. Again, it is a simple function which you may complicate even more

5. List gn stands for the set of conductances connected to the - input of the OA, excepting Rf. If no resistance is connected, then $gn = \{0\}$.

6. List gp for the resistances connected to the + terminal. If no resistance is connected, $gp = \{\}$.

7. List vn is the set of inputs in the inverting side. If no input is present, then $vn = \{\}$ or $vn = \{0\}$.

8. List vp is the set of inputs in the non verting side. If no input is present, then $vp = \{\}$ or $vp = \{0\}$.

9. Rf must have a finite value, including 0.

Fig. 5.31 on the following page shows some circuits with operational amplifiers, each one with the description of the addsub() function. Observe that vp and gp must be of the same dimension, so a dummy 0 is included to vp when a resistance is grounded. Try the examples in your calculator..

5.6 Closing Remarks

As you might have imagined by now, it is possible to stay with reduction and transformation methods for the rest of the book. These methods require from you to develop your intuition, which is excellent in many ways, both for your professional development and your improvement in analysis and tools manipulating skills. Literature is full of such methods; you are encouraged to look at them.

As a suggestion, develop a library of functions or programs, specially for those cases which appear often, so you can boost your speed in solving problems. But write the functions yourself, don't look for them at the internet, unless you already know how to do it. Doing it yourself you don't miss the joy of learning and discovering! To help you better in this task, don't forget to develop everything within one document in your TI-nspire cx.

Finally, it is important to realize that steps and procedures depend on your intuition and, to develop the intuition you must practice, practice and practice. But don't forget: practice paying attention to procedure! The calculator won't teach you!

90   CHAPTER 5. TRANSFORMATIONS AND REDUCTIONS

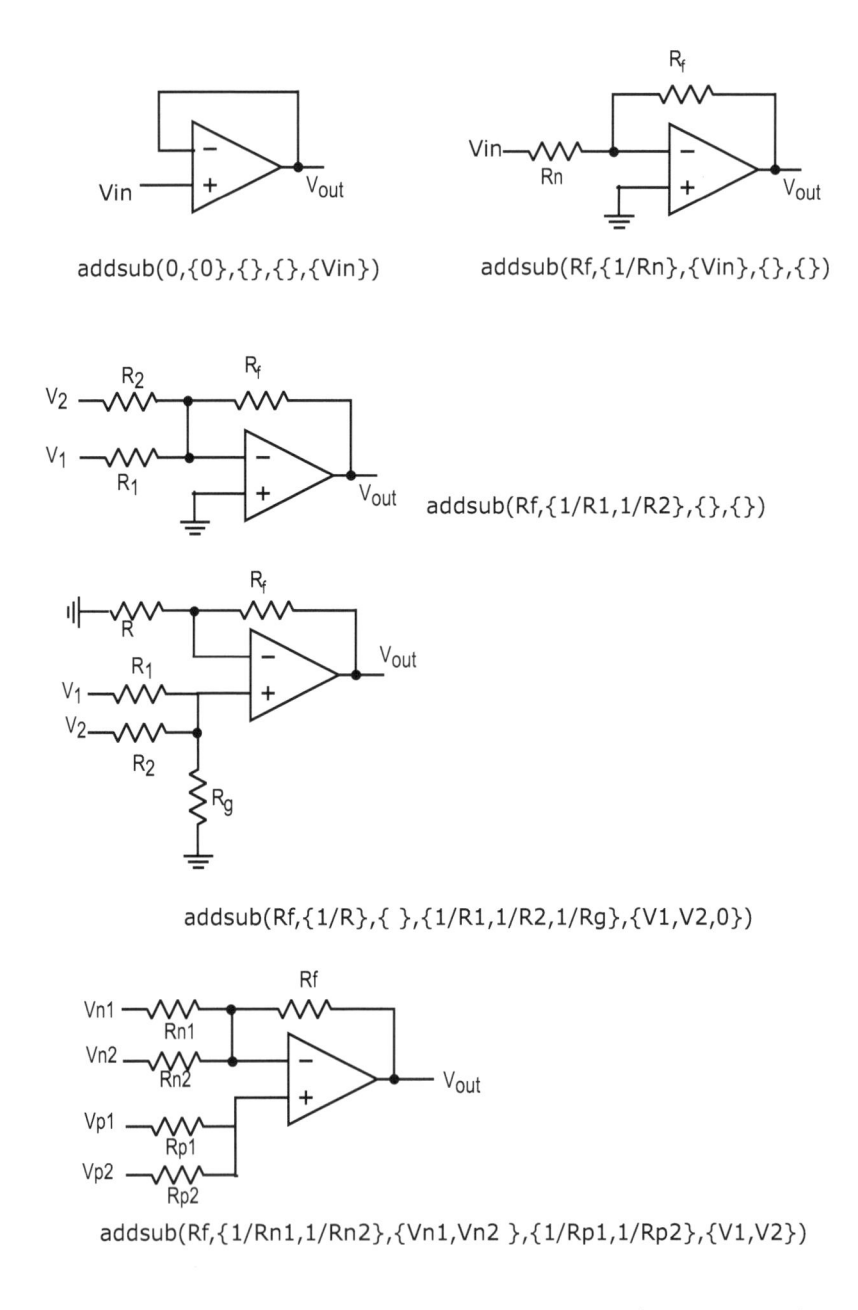

addsub(0,{0},{},{},{Vin})

addsub(Rf,{1/Rn},{Vin},{},{})

addsub(Rf,{1/R1,1/R2},{},{})

addsub(Rf,{1/R},{ },{1/R1,1/R2,1/Rg},{V1,V2,0})

addsub(Rf,{1/Rn1,1/Rn2},{Vn1,Vn2 },{1/Rp1,1/Rp2},{V1,V2})

Figure 5.31: Examples to use with function addsub(rf,gn,vn,gp,vp)

CHAPTER 6

Nodal Analysis

Nodal analysis is one of the most popular analysis methods based on a systematic set up of equations. In this chapter I review effective ways to write and solve these equations, and include programs. I assume that the reader can use matrices to solve linear equations.

6.1 Introduction to nodal analysis

Nodal analysis uses the concept of *node potential*, which is the voltage difference between a node k and a reference node called *ground* which, by definition, has potential of 0 V. Any other voltage can be expressed in terms of these potentials, as illustrated in Fig. 6.1. A circuit with N nodes will have N-1 node potentials.

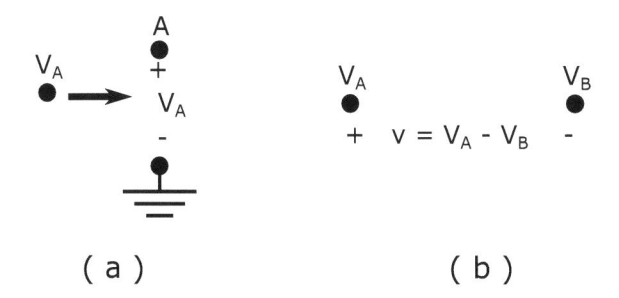

Figure 6.1: a) The concept of node potential, b) voltage between two nodes

Mathematically, any node can be chosen as ground. If you have already found the potentials for a circuit and decide to change ground to another node B, then all potentials can be adjusted to the new reference by subtracting the original potential V_B, as illustrated in Fig. 6.2. Therefore, do not worry too much if you selected one node or another as reference. However, pay attention to the sign of voltages.

Remark: For mathematical analysis, selection of ground is a matter of convenience. Physical ground is quite different, since many parasitics should be considered. This is particularly important in many cases, such as layouts of printed circuit boards, design of integrated circuits, and many other instances.

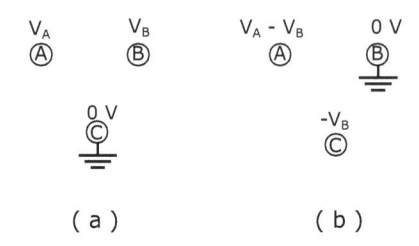

Figure 6.2: Changing reference and effect on potentials: a) original, b) after change

The nodal method consists in using the node potentials as main variables of a set of equations, called *nodal equations*, which are essentially those derived from Kirchhoff's Current Law. The complete set may also include currents not dependent on voltages, making it necessary to add other equations or constraints. Once the system is solved for the node potentials and those currents, any other circuit variable can be calculated in terms of those results.

Since each nodal equation is actually a current equation based on Kirchhof's current law, let us express this principle as follows for our goals:

The sum of unknown currents leaving a node is equal to the sum of known currents entering to the node.

The unknown currents leaving the node can be divided in three groups, which are mutually independent:

A. Currents leaving through resistances

B. Currents not in group A which are dependent on voltages, like those in voltage controlled current sources.

C. Currents independent of voltages, such as those in voltage sources or at the output of an ideal operational amplifier.

For each current present in group C, we need another equation (element equation) or else a condition (like a potential being known).

Currents from groups A and B will generate an expression of the form

6.2. ONLY RESISTANCES AND CURRENT SOURCES   93

$$Y_{k1} V_1 + Y_{k2} V_2 + Y_{k3} V_3 + \cdots + Y_{k(N-1)} V_{N-1} \quad (6.1)$$

which does not depend on currents from group C.

When no calculators were available, it was of outmost importance to have as few equations as possible. For that reason, methods to avoid including currents from group C were introduced, and are still being presented in textbooks. This is a good practice, since sometimes it becomes really necessaryor convenient to work with a reduced number of equations. It is interesting to remark that this practice is still so important that analysis where group C is included is called *Modified Nodal Analysis*!

I present the nodal analysis method in two steps. First, I follow the traditional textbook approach of just working with resistances, known currents and currents dependent on voltages. The method is quite direct and easily programmable. I present a program example for it.

To avoid the presence of voltage unrelated currents at this stage, any voltage source may be converted into current source, as explained in section 5.4. Another way to avoid these currents without using transformations is the "supernode" method[1] is applied. This method can also be written by inspection.

In the second part, I present the modified nodal analysis (MNA), that is, I include unknown currents not dependent on voltages. This is avoided in textbooks which are oriented to hand analysis. Yet, we work here with the calculator! No need to avoid complexity! Besides, MNA will allow us to apply our tools faster and easier. MNA is also programmable.

6.2 Only resistances and current sources

Take a circuit with $n + 1$ nodes, including ground. The circuit contains only "known" independent current sources, resistances and voltage dependent current sources. Our objective is to write by inspection the equation at each node m different from ground in the form

$$Y_{m1} V_1 + Y_{m2} V_2 + \cdots + Y_{mm} V_m + \cdots + Y_{mn} V_n = In_m \quad (6.2)$$

The left hand side of (6.2) represents, after algebraic manipulation, the sum of unknown currents leaving the node. The right hand side is the sum of currents entering to the node. In this equation:

- $V_1, V_2, \ldots$ are the *node potentials* for nodes 1, 2, ...
- Coefficients Y_{km} are called *nodal admittances* or *nodal conductances*
- In_m is called the *nodal current* at node m.

---
[1]The correct name for this method should be "cut-set" method. However, "supernode" is already accepted by the community.

For convenience, voltages and currents of the elements will be written with small letters, while potentials and node currents with capital letters. When this convention has not been followed, it must be clear from context what is the meaning of the variable.

The set of these nodal equations can be written in matrix form as

$$\mathbf{Yn\,Vn = In} \qquad (6.3)$$

6.2.1 Theoretical principles

If you already know to write nodal equations of the form (6.2) by inspection you may skip this part.

The basis for node equations is Kirchhoff's Current Law, which states that the sum of currents leaving a node is equal to the sum of currents entering the node. In Fig. 6.3, let us consider node m, with the following

CONVENTION: currents leaving from the node are those through resistances and voltage-dependent current sources, and entering are known currents from independent sources.

It follows from this convention that

I_m is the sum of known currents entering node m: $I_m = \sum i_j$ (6.4)

Here, any known current leaving the node is taken with negative sign. Consider now the isolated node m shown in Fig. 6.3.

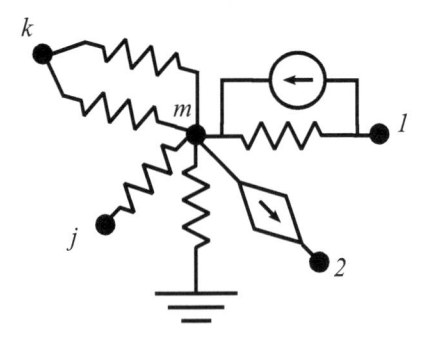

Figure 6.3: Isolated node m in circuit.

The nodal admittances Y_{mj} in (6.2) consists of two components, independent of each other. One is G_{mj} due to resistances, and another one go_{mj} arising from voltage-dependent current sources. That is,

$$Y_{mj} = G_{mj} + go_{mj} \qquad (6.5)$$

6.2. ONLY RESISTANCES AND CURRENT SOURCES

Component of G_{mj} of Y_{mj} due to resistances.

Looking at Fig. 6.4, we can state the following:

> *For each resistance R shared by node m and node j, there will be a current leaving the node in the form*
> $$\frac{V_m - V_j}{R} = G(V_m - V_j) = GV_m - GV_j \quad (6.6a)$$
>
> *And for each resistance connected between m and ground there is a current of the form*
> $$\frac{V_m}{R} = GV_m \quad (6.6b)$$
>
> *Here, $G = 1/R$ is the conductance of the resistance R.*

Hence, we can conclude that when regrouping terms in the equation of node m, the coefficient of V_m will be the sum of all conductances connected to node m, while the coefficient of V_j for $j \neq m$ will be the negative of the sum of conductances connecting node m to node j, that is, the conductances shared by both nodes. Mathematically, this is expressed by

$$G_{mj} = \begin{cases} +\sum \text{Conductances connected to node } m & \text{for } j = m \\ -\sum \text{Conductances shared by nodes } m \text{ and } j & \text{for } j \neq m \end{cases} \quad (6.7)$$

Referring to the equation in matrix form, Fig. 6.4 shows how a resistance at node m contribute to the equation, or to be specific, to the equation row in the matrix. Labels over the row show the columns and respective variables.

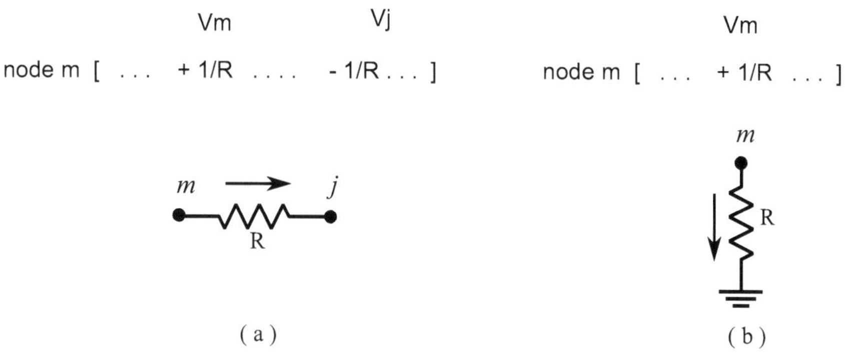

```
              Vm            Vj                                  Vm
node m [ ... + 1/R  ....  - 1/R ... ]       node m [  ...    + 1/R   ... ]
```

(a)   (b)

Figure 6.4: Currents leaving node m through resistances.

Voltage-controlled current sources

Unknown currents leaving from nodes include those through voltage controlled current sources of the type $go\,Vx = go(V_p - V_q)$. Since each source leaves from a node m and enters a node h, only coefficients in the equations at those two nodes are affected. Assuming connections as illustrated in Fig. 6.5, we write the equations at nodes m and h as

$$\overbrace{\begin{matrix}G_{m1}V_1 + \cdots + G_{mp}V_p + \cdots + G_{mq}V_q + \cdots \\ G_{h1}V_1 + \cdots + G_{hp}V_p + \cdots + G_{hq}V_q + \cdots\end{matrix}}^{\text{Component of R's}}\begin{matrix}+g_o(V_p - V_q) = In_m \\ -g_o(V_p - V_q) = In_h\end{matrix}$$
$$\underbrace{\phantom{G_{h1}V_1 + \cdots + G_{hp}V_p + \cdots + G_{hq}V_q + \cdots}}_{\text{Component of R's}}$$

After some Algebra, we arrive at

Node m: $G_{m1}\,V_1 + \cdots + (G_{mp}+g_o)\,V_p + \cdots + (G_{mq}-g_o)\,V_q + \cdots = In_m$ (6.8a)

Node h: $G_{h1}\,V_1 + \cdots + (G_{hp}-g_o)\,V_p + \cdots + (G_{hq}+g_o)V_q + \cdots = In_h$ (6.8b)

```
                Vp              Vq                          Vp              Vq
node m [ ... Gmp + go  ...  Gmq - go ... ]    node m [ ... Gmp + (a/R) ... Gmq - (a/R) ... ]
node h [ ... Ghp - go  ...  Ghq + go ... ]    node h [ ... Ghp - (a/R) ... Ghq + (a/R) ... ]
```

(a)   (b)

Figure 6.5: Voltage controlled current source and nodal coefficients.

Fig. 6.5 also illustrates how the source is inserted into the equations. Notice that the current controlled source was converted to voltage controlled source using Ohm's law. If you have problems remembering this scheme, then write down the equations first with the controlled sources separated from resistances and rearrange coefficients as explained above.

6.2.2 Rule to generate the equations

For convenience, let us repeat the rules to write by inspection the set of equations for the case when the circuit consists of only independent current sources and resistances.

At each node m, the nodal equation is written using the following rules:

$$I_m \text{ is the sum of known currents entering node } m: I_m = \sum i_j \quad (6.9)$$

A current source entering the node is positive, and negative if it leaves the node.

$$G_{mj} = \begin{cases} +\sum \text{Conductances connected to node } m & \text{for } j = m \\ -\sum \text{Conductances shared by nodes } m \text{ and } j & \text{for } j \neq m \end{cases} \quad (6.10)$$

Two examples applying (6.4) and (6.10) will illustrate the algorithm. Equations are written in matrix form. The first one goes step by step, showing clearly the application of the criteria. The second one is straightforward, but the reader is encouraged to verify each matrix component.

Example 6.1 *A step by step example, Fist Part:*

Let us start with step-by-step setting of the nodal equations for the circuit in Fig. 6.6. Nodes are labeled 1, 2 and 3. We write the matrix $\mathbf{Y_n}$ and vector $\mathbf{I_n}$ by inspection using (6.10) and (6.9).

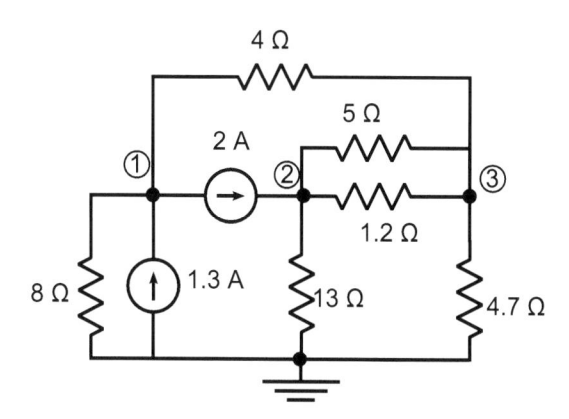

Figure 6.6: Circuit to analyze with nodal method.

Figures 6.7, 6.8, and 6.9 show how the components of the matrices are found. The process is illustrated by taking subcircuits focused on the individual nodes 1, 2, and 3, keeping the components connected to them and seeing

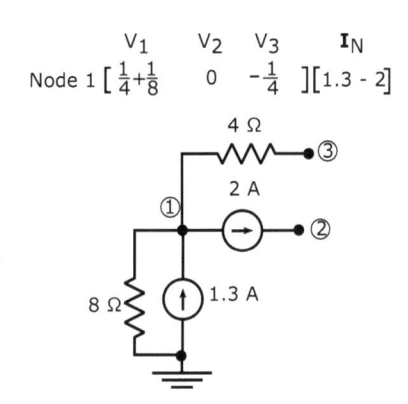

$$\text{Node 1}\begin{bmatrix} V_1 & V_2 & V_3 & \mathbf{I_N} \\ \tfrac{1}{4}+\tfrac{1}{8} & 0 & -\tfrac{1}{4} \end{bmatrix}\begin{bmatrix} 1.3 - 2 \end{bmatrix}$$

Figure 6.7: Forming matrix and equations for node 1.

how those elements are shared with other nodes. Above each subcircuit, the row of "coefficients" in the equation of the respective node is displayed.

Consider node 1, at Fig. 6.7:

1. *Two resistances are connected to node 1: one of 8 Ω and another one of 4 Ω. Hence, the coefficient of V_1 in this equation is $(1/4 + 1/8)$; this is the element shown in the row vector of $\mathbf{Y_n}$ above the circuit, just under the label V1.*

2. *There are no resistances shared by nodes 1 and 2, so the coefficient of V_2 is 0, as shown under the label V2.*

3. *The resistance of 4 Ω is shared by nodes 1 and 3, so the coefficient of V_3 is -1/4.*

4. *For the respective row in the vector $\mathbf{I_n}$, a current of 1.3 A is entering the node and another one of 2 A leaving the node. The element in the row of node 1 is therefore (1.3 - 2).*

Look now at node 2, Fig. 6.8:

1. *There are no resistances shared by nodes 1 and 2, so the coefficient of V_1 is 0, as shown under the label V1.*

2. *The three resistances connected to node 2 are of 5 Ω, 1.2 Ω, and 13 Ω. Hence, the coefficient of V_2 in this equation is $(1/5 + 1/1.2 + 1/13)$; this is the element shown in the row vector of $\mathbf{Y_n}$ above the circuit, just under the label V2.*

3. *The resistances of 5 Ω and 1.2 Ω are shared by nodes 2 and 3, so the coefficient of V_3 is -1/5 - 1/1.2.*

6.2. ONLY RESISTANCES AND CURRENT SOURCES 99

$$\text{Node 2 } \begin{bmatrix} V_1 & V_2 & V_3 & I_N \\ 0 & \frac{1}{5}+\frac{1}{1.2}+\frac{1}{13} & -\frac{1}{5}-\frac{1}{1.2} \end{bmatrix} \begin{bmatrix} 2 \end{bmatrix}$$

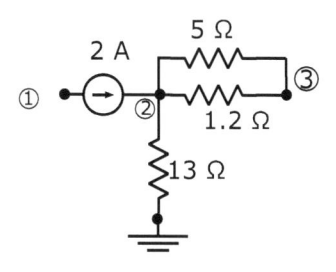

Figure 6.8: Forming matrix and equations for node 2.

4. Finally, for the row in the vector $\mathbf{I_n}$, we see the current source of 2 A entering into the node. Thus, in the row of node 2 we have 2.

It should be clear that the coefficient of V_k in the equation of node m is the same as the coefficient of V_m in the equation of node k, since both are defined by the resistances shared by both nodes. This can be checked in the two steps above.

The reader is encouraged to find the equation for node 3 looking at Fig. 6.9.

$$\text{Node 3 } \begin{bmatrix} V_1 & V_2 & V_3 & I_N \\ -\frac{1}{4} & -\frac{1}{5}-\frac{1}{1.2} & \frac{1}{4}+\frac{1}{5}+\frac{1}{1.2}+\frac{1}{4.7} \end{bmatrix} \begin{bmatrix} 0 \end{bmatrix}$$

Figure 6.9: Forming matrix and equations for node 3.

The set of equations can now be summarized in matrix form as

$$\begin{array}{c} \begin{array}{ccc} V_1 & V_2 & V_3 \end{array} In \\ \begin{array}{c} Node1 \\ Node2 \\ Node3 \end{array} \left[\begin{array}{ccc} \frac{1}{8}+\frac{1}{4} & 0 & -\frac{1}{4} \\ 0 & \frac{1}{5}+\frac{1}{1.2}+\frac{1}{13} & -\frac{1}{5}-\frac{1}{1.2} \\ -\frac{1}{4} & -\frac{1}{5}-\frac{1}{1.2} & \frac{1}{4}+\frac{1}{5}+\frac{1}{1.2}+\frac{1}{4.7} \end{array} \right] \left[\begin{array}{c} V_1 \\ V_2 \\ V_3 \end{array} \right] = \left[\begin{array}{c} 1.3-2 \\ 2 \\ 0 \end{array} \right] \end{array}$$
(6.11)

The solution is $[V_1\ V_2\ V_3]^T = [0.662, 5.332, 3.793]^T$

Let us now introduce this example onto the calculator, and solve.

Example 6.2 *A step by step example, Second Part:*

We introduce here the data for the previous example onto the calculator, illustrating the steps of Figs. 6.7 to 6.9. In your calculator enter
yn:= [menu] 7 1 1 [del] 3 [tab] [del] 3 [enter] *to start your matrix.*

Now enter the rows. First row:

1/4+1/8 [tab] 0 [tab] [(-)] 1/4 [tab]

Continue with the second row:

0 [tab] 1/5 + 1/1.2 + 1/13 [tab] [(-)] 1/5 - 1/1.2 [tab]

and then the third row.

[(-)] 1/4 [tab] [(-)] 1/5 - 1/1.2 [tab] 1/5 + 1/1.2 + 1/4 + 1/4.7 [enter]

to display

$$\left[\begin{array}{ccc} .375 & 0.000 & -.250 \\ 0.000 & 1.110 & -1.033 \\ -.250 & -1.033 & 1.496 \end{array} \right]$$

Next, create your 3x1 matrix **In** *with*
in:= [menu] 7 1 1 [del] 3 [tab] [del] 1 [enter] *followed by*
1.3 -2 [tab] 2 [tab] 0 [enter]

You can now solve with simul(yn,in) *or*

$$\text{yn}^{-1}\ \boxed{\times}\ \text{in}\ \boxed{\text{enter}} \rightarrow \left[\begin{array}{c} .662 \\ 5.332 \\ 3.793 \end{array} \right]$$

6.2. ONLY RESISTANCES AND CURRENT SOURCES

If you prefer to use the augmented matrix, creat matrix y *of order 3×4 with the rows*

First row:

1/4+1/8 [tab] 0 [tab] [(-)] 1/4 [tab] 1.3 -2 [tab]

Second row:

0 [tab] 1/5 + 1/1.2 + 1/13 [tab] [(-)] 1/5 - 1/1.2 [tab] 2 [tab]

And third row:

[(-)] 1/4 [tab] [(-)] 1/5 - 1/1.2 [tab] 1/5+1/1.2+1/4+1/4.7 [tab] 0 [enter]

Your matrix y *must now be displayed as*

$$\begin{bmatrix} .375 & 0.000 & -.250 & -.700 \\ 0.000 & 1.110 & -1.033 & 2.000 \\ -.250 & -1.033 & 1.496 & .000 \end{bmatrix}$$

Now apply the transformation Reduced Row-Echelon Form rref()*, with* [menu] 7 5 *and look for the solution at the last column:*

rref(y) [enter] → $\begin{bmatrix} 1.000 & 0.000 & 0.000 & 0.662 \\ 0.000 & 1.000 & 0.000 & 5.332 \\ 0.000 & 0.000 & 1.000 & 3.793 \end{bmatrix}$

The following example refers also to a circuit containing only resistances and independent current sources. Nodal equations are used again, but this time we write the nodal equations directly by inspection, following the previous rules, and use the results from the solution of the set of equations. If you have problems doing the first part, work the problem using an approach similar to example 6.1, where the individual sub circuits were isolated for easier identification.

Example 6.3 *For the circuit of Fig. 6.10, find Io and the power delivered by the current sources. Use nodal equations in the process.*

Solution: *Before proceeding to set up equations, first our strategy. Since our equations solve for node potentials, we need to relate these potentials with the requested results.*

A generated power by a source is found by multiplying the voltage and current, with the current leaving the + side of voltage. On the other hand,

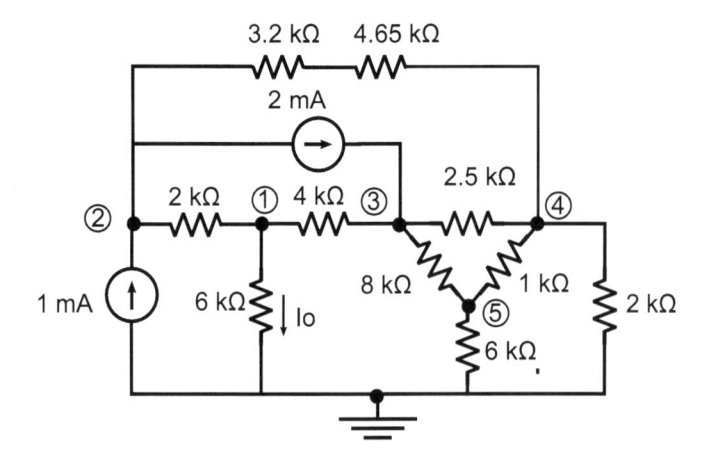

Figure 6.10: Another circuit to analyze with nodal method.

Io is the current in a resistance. Hence, the relationship between the node potentials and the requested results is as follows. For the generated powers:

$$\text{generated by the 1 mA source: } P_1 = (1\,\text{mA})\,V_2$$

$$\text{generated by the 2 mA source: } P_2 = (2\,\text{mA})\,(V_3 - V_2)$$

For the current Io,

$$io = \frac{V_1}{6000}\,(\text{A})$$

Now, there is no special request concerning the two resistances in series between nodes 2 and 4, so we can combine them in one equivalent resistance. Don't do it by yourself, just enter the expression in the formula.

That said, we are looking to solve for $\mathbf{Vn} = [V_1, V_2, \ldots, V_5]^T$ *in*

$$\mathbf{Yn\,Vn} = \mathbf{In} \tag{6.12}$$

where $\mathbf{Yn}$ *is (with labels attached for easy reference)*

$$\begin{array}{c} & V_1 & V_2 & V_3 \\ N.1 \\ N.2 \\ N.3 \\ N.4 \\ N.5 \end{array} \left[\begin{array}{ccc} \frac{1}{6\text{ E3}} + \frac{1}{4\text{ E3}} + \frac{1}{2\text{ E3}} & -\frac{1}{2\text{ E3}} & -\frac{1}{4\text{ E3}} \\ -\frac{1}{2\text{ E3}} & \frac{1}{2\text{ E3}} + \frac{1}{3.2\text{ E3}+4.65\text{ E3}} & 0 \\ -\frac{1}{4\text{ E3}} & 0 & \frac{1}{8\text{ E3}} + \frac{1}{2.5\text{ E3}} + \frac{1}{4\text{ E3}} \\ 0 & -\frac{1}{3.2\text{ E3}+4.65\text{ E3}} & -\frac{1}{2.5\text{ E3}} \\ 0 & 0 & -\frac{1}{8\text{ E3}} \end{array} \right.$$

6.2. ONLY RESISTANCES AND CURRENT SOURCES

$$\begin{array}{cccc}
& \cdots & V4 & V5 \\
(N.1) & \cdots & 0 & 0 \\
(N.2) & \cdots & -\frac{1}{3.2\text{ E}3+4.65\text{ E}3} & 0 \\
(N.3) & \cdots & -\frac{1}{2.5\text{ E}3} & -\frac{1}{8\text{ E}3} \\
(N.4) & \cdots & \frac{1}{2.5\text{ E}3}+\frac{1}{3.2\text{ E}3+4.65\text{ E}3}+\frac{1}{2\text{ E}3}+\frac{1}{1\text{ E}3} & -\frac{1}{1\text{ E}3} \\
(N.5) & \cdots & -\frac{1}{1\text{ E}3} & \frac{1}{1\text{ E}3}+\frac{1}{6\text{ E}3}+\frac{1}{8\text{ E}3}
\end{array}$$

and

$$\mathbf{In} = \begin{bmatrix} 0 \\ 1\text{E}\boxed{(-)}3 - 2\text{E}\boxed{(-)}3 \\ 2\text{E}\boxed{(-)}3 \\ 0 \\ 0 \end{bmatrix}$$

Now enter the matrices in the calculator and store them to `yn` *and* `in`, *respectively. Store the solution to* `vn`.

Use `simult(yn,in)` `sto→` `vn` *or matrix multiplication. With three decimal figures, we have*

$$\mathtt{yn}^{-1} * \mathtt{in} \quad \boxed{\text{sto}\rightarrow} \quad \mathtt{vn} \rightarrow \begin{bmatrix} 0.501 \\ -0.918 \\ 3.673 \\ 1.363 \\ 1.411 \end{bmatrix}$$

The elements in the vector `vn` *are, respectively,* $V_1, V_2 \ldots V_5$. *We can therefore find the desired answers for this problem using these vector components as follows:*

For P_1 *and* P_2, *respectively,*

1 `EE` `(-)` 3 × vn[2,1] `enter` → -917.806E⁻6

2 `EE` `(-)` 3 × (vn[3,1] - vn[2,1]) `enter` → 9.182E⁻3

Thus, the 1 mA source is actually absorbing 918 μW *while the 2 mA source is generating* 9.18 mW.

For current I_o: vn[1,1] / 6000 → 83.526E⁻6.
Hence, this current is 83.526 μA.

6.2.3 Scaling units

If both sides of a node equation are multiplied by 1000, the results are unaffected. This means that we could enter the resistances in kΩ units and the current sources in mA units, without changing the values or units of the voltages (V=RI). This brief remark is useful for us, both in time saving and in numerical rounding, since numbers are not as small.

With this remark, the current vector **In** in the previous example can be entered as

$$\mathbf{In} = [0,\ 1-2,\ 2,\ 0,\ 0]^T$$

and the matrix **Yn** as

$$
\begin{array}{c}
\\
Node1 \\
Node2 \\
Node3 \\
Node4 \\
Node5
\end{array}
\left[
\begin{array}{ccc}
V_1 & V_2 & V_3 \\
\frac{1}{6}+\frac{1}{4}+\frac{1}{2} & -\frac{1}{2} & -\frac{1}{4} \\
-\frac{1}{2} & \frac{1}{2}+\frac{1}{3.2+4.65} & 0 \\
-\frac{1}{4} & 0 & \frac{1}{8}+\frac{1}{2.5}+\frac{1}{4} \\
0 & -\frac{1}{3.2+4.65} & -\frac{1}{2.5} \\
0 & 0 & -\frac{1}{8}
\end{array}
\right.
$$

$$
\begin{array}{c}
Node1 \\
Node2 \\
Node3 \\
Node4 \\
Node5
\end{array}
\begin{array}{cc}
\cdots & V4 & V5 \\
\cdots & 0 & 0 \\
\cdots & -\frac{1}{3.2+4.65} & 0 \\
\cdots & -\frac{1}{2.5} & -\frac{1}{8} \\
\cdots & \frac{1}{2.5}+\frac{1}{3.2+4.65}+\frac{1}{2}+\frac{1}{1} & -\frac{1}{1} \\
\cdots & -\frac{1}{1} & \frac{1}{1}+\frac{1}{6}+\frac{1}{8}
\end{array}
\Bigg]
$$

Smaller matrices, easier and faster to enter. Notice that if a conductance value is already entered in the matrix, it should be in mS units.

6.2.4 An example with dependent sources

Let us work an example with dependent sources to illustrate dealing with them. We take advantage of the example to also illustrate again the advantages of scaling units.

Example 6.4 *For the circuit in Fig. 6.11, write down the matrices* **Yn** *and* **In** *of the nodal equation* **Yn Vn = In**. *Apply scaling by* 10^3, *i. e., resistance units in kilohms, currents in mA and conductance units in mS.*

Looking at the circuit, we see that there are two controlled current sources. Although the first one is current controlled, we convert it to voltage controlled

6.2. ONLY RESISTANCES AND CURRENT SOURCES

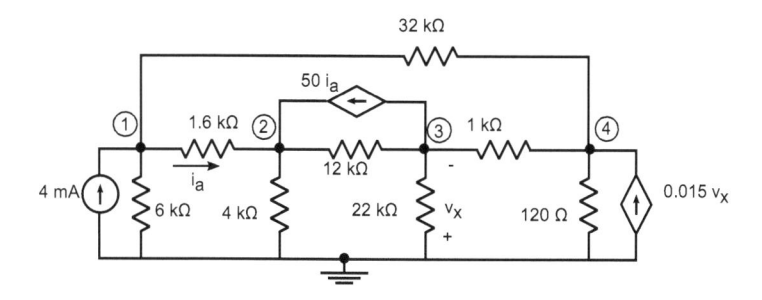

Figure 6.11: Voltage controlled current source and nodal coefficients.

using the fact that $i_a = (V_1 - V_2)/1.6k\Omega$. Hence, after applying scaling, this source adds
$$\frac{50}{1.6}(V_1 - V_2)$$
to the equation at node 3, and subtracts it at the equation of node 2. Coefficients to modify are Y_{13}, Y_{23}, Y_{12} and Y_{22}. Therefore

in equation of node 3: add $\frac{50}{1.6}$ to coefficient of V_1 and subtract $\frac{50}{1.6}$ from coefficient of V_2

in equation of node 2: subtract $\frac{50}{1.6}$ from coefficient of V_1 and add $\frac{50}{1.6}$ to coefficient of V_2

On the other hand, the voltage controlled current source $0.015\,v_x = -0.015V_3$ will be subtracted at equation 4, affecting Y_{43} only, where we add 15 mS.

We can now work the matrix $\mathbf{Yn}$ directly as shown next, with labels attached for easy reference, and applying scaling,

$$\begin{array}{c} \\ N1 \\ N2 \\ N3 \\ N4 \end{array} \begin{array}{c} V_1 \qquad\qquad V_2 \qquad\qquad V_3 \qquad\qquad V_4 \\ \left[\begin{array}{cccc} \frac{1}{6}+\frac{1}{1.6}+\frac{1}{32} & -\frac{1}{1.6} & 0 & -\frac{1}{32} \\ -\frac{1}{1.6}-\frac{50}{1.6} & \frac{1}{1.6}+\frac{1}{4}+\frac{1}{12}+\frac{50}{1.6} & -\frac{1}{12} & 0 \\ 0+\frac{50}{1.6} & -\frac{1}{12}-\frac{50}{1.6} & \frac{1}{12}+\frac{1}{1}+\frac{1}{22} & -\frac{1}{1} \\ -\frac{1}{32} & 0 & -\frac{1}{1}+\mathbf{15} & \frac{1}{1}+\frac{1}{0.120}+\frac{1}{32} \end{array} \right] \end{array}$$
(6.13)

Here, bold entries are the ones due to the controlled sources.

Vector $\mathbf{In}$ is
$$\mathbf{In} = [4\ 0\ 0\ 0]^T$$

The result for this circuit is $[14.3, -2.8, -204.2, 305.3]^T$, whose values are the node potentials in Volt (V) units.

6.2.5 Modifying a circuit

Assume that *after* you have already set up the circuit equations, elements are added to, or deleted from, the circuit, without altering the number of nodes. That means that new elements are added connecting two nodes or a node to ground, in parallel to existing elements or else introducing one where there was none.

Figs. 6.4 and 6.5 illustrate how can we proceed without need to reintroduce matrices from the start. You can do it in the matrix editor, or directly on the command line. The rule is as follows:

Assume that you have already produced a matrix **Yn** *for your circuit, and you connect a new resistance* R *between nodes k and j. Then you can obtain the matrix for the new circuit with the following edits:*

$$\text{Yn[k,k]} + 1/\text{R} \boxed{\text{sto}\rightarrow} \text{Yn[k,k]} \tag{6.14a}$$

$$\text{Yn[j,j]} + 1/\text{R} \boxed{\text{sto}\rightarrow} \text{Yn[j,j]} \tag{6.14b}$$

$$\text{Yn[k,j]} - 1/\text{R} \boxed{\text{sto}\rightarrow} \text{Yn[k,j]} \tag{6.14c}$$

$$\text{Yn[j,k]} - 1/\text{R} \boxed{\text{sto}\rightarrow} \text{Yn[j,k]} \tag{6.14d}$$

The above equations are applied whenever a new resistance is connected. However, we can work similarly other situations.

1. If the resistance is removed or disconnected, the signs for $1/R$ are changed in all equations (6.14)

2. If the resistance is connected between node k and ground, apply only (6.14a)

3. If you find you have entered a wrong resistance R value, then "disconnect" the mistaken $1/R$ and "connect" the correct one in just one step in the equations.

These equations and remarks assume that no controlled source is affected with the modification. If there is any, study and change the corresponding entries.

The following exercise invites the reader to apply the above procedures. For this example, I recommend the reader to draw the modified schematics of the new circuit.

6.2. ONLY RESISTANCES AND CURRENT SOURCES

Example 6.5 *In example 6.4, the matrix* $\mathbf{Y}n$ *calculated in (6.13) is stored as* Y *in the calculator, while the source vector as* Z. *Displayed in engineering mode,* $\mathbf{Y}$ *is*

$$\begin{bmatrix} 822.92\text{E-}3 & -625.\text{E-}3 & 0.\text{E}0 & -31.25\text{E-}3 \\ 5.2083\text{E}0 & 32.208\text{E}0 & -83.333\text{E-}3 & 0\text{E}0 \\ 31.25\text{E}0 & -31.333\text{E}0 & 1.1288\text{E}0 & -1.\text{E}0 \\ -31.25\text{E-}3 & 0\text{E}0 & 14.\text{E}0 & 9.3646\text{E}0 \end{bmatrix}$$

Remember that all values are in mS *units since resistances were entered in* kΩ *units and conductances in* mS *units!*

Find the new $\mathbf{Y}n$ *if the following modifications are made*

1. A new resistance of 3 kΩ is connected in parallel with the 1 kΩ resistance, between nodes 3 and 4, and the resistance in parallel with the independent source is erased.

2. The 120 Ω resistance value is changed to 12.3 kΩ.

To take into account the introduction of the 3 kΩ *resistor:*

Y[3,3] + 1/3 sto→ Y[3,3]
Y[4,4]+ 1/3 sto→ Y[4,4]
Y[4,3]- 1/3 sto→ Y[4,3]
Y[3,4]- 1/3 sto→ Y[3,4]

To suppress the resistance of 6 kΩ *in parallel with the independent source, between node 1 and ground:*

Y[1,1]- 1/6 sto→ Y[1,1]

Finally, to modify the 120 Ω *resistance, taking into account the scaling:*

Y[4,4]- 1/0.120 + 1/12.3 sto→ Y[4,4]

6.2.6 Working voltage sources with source transformation

In circuits with voltage sources we can apply source transformation and the shifting theorem explained in section 5.4 to get a circuit containing only current sources. The transformations eliminate all information for the currents at the voltage sources, and reduce the number of nodes and, as a consequence, the number of equations. If you still need potentials or currents that have been eliminated, you must add additional equations. Working an example may be the best way to understand the method.

Example 6.6 *Let us work the circuit of example Fig. 6.12(a). When you apply source transformation to the independent source and the shifting theorem to the dependent source, followed by source transformations, you arrive at Fig. 6.12(b)[2].*

(a)

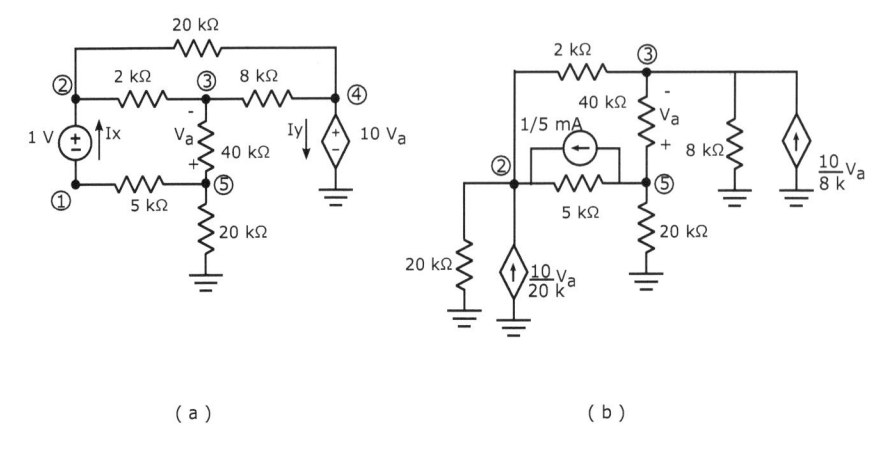

(b)

Figure 6.12: Analysis of circuit of Fig. 6.18 with current sources: (a) original circuit; (b) transformed circuit.

Observe that in the transformed circuit we have lost nodes 1 and 5, as well as currents I_x and I_y. All we need to solve now are three equations.

In the transformed circuit, using scaling for resistances in kΩ units, the equations are

$$\begin{array}{c} \\ N.2 \\ N.3 \\ N.5 \end{array} \begin{array}{c} V_2 \quad\quad\quad V_3 \quad\quad\quad V_5 \\ \begin{bmatrix} \frac{1}{2}+\frac{1}{5}+\frac{1}{20} & -\frac{1}{2}+\frac{10}{20} & -\frac{1}{5}-\frac{10}{20} \\ -\frac{1}{2} & \frac{1}{2}+\frac{1}{40}+\frac{1}{8}+\frac{10}{8} & -\frac{1}{40}-\frac{10}{8} \\ -\frac{1}{5} & -\frac{1}{40} & \frac{1}{40}+\frac{1}{20}+\frac{1}{5} \end{bmatrix} \begin{bmatrix} V_2 \\ V_3 \\ V_5 \end{bmatrix} = \begin{array}{c} In \\ \begin{bmatrix} \frac{1}{5} \\ 0 \\ -\frac{1}{5} \end{bmatrix} \end{array} \end{array}$$

The result for this system is $[V_2 \ V_3 \ V_5]^T = [-1.801 \ -1.96 \ -2.215]^T$. With
$$V_2 - V_1 = 1 \quad \text{and} \quad V_4 = 10(V_5 - V_3)$$
we can get the remaining potentials. Similarly, equations at nodes 1 and 4 allow us to get the currents.

---

[2]Refer to section 5.4 if you need to refresh these concepts

6.3 Other considerations for nodal equations

Before going on to other situations for node equations, let us introduce concepts and special applications. Even though introduced here, they can be applied to any later situation where known sources are not current sources.

6.3.1 Indefinite admittance matrix

When ground is not connected to any element of the circuit, the resultant matrix **Yn** is called *indefinite admittance matrix* (IAM). By extension, **In**, may be called indefinite nodal current vector. Both matrices **Yn** and **In** have the characteristic that the sum of the rows and columns are 0. As a consequence, the determinant of **Yn** is 0 an the set of equations does not have a unique solution. Yet, the IAM has many uses[3]. For this book we are interested in the following property.

*When one of the nodes in the circuit is connected to ground, the corresponding column and row in the indefinite **Yn** and **In** are deleted, and the matrix and vector become the definite nodal admittance matrix and vector that we already know.*

We will use this property in programming, but many other situations may be considered for application

6.3.2 Superposition in Nodal Analysis

When we have several sources, we can separate their roles to apply superposition in the same way that was explained in subsection 4.3 and write the nodal equations in the form (4.5) as

$$\mathbf{Yn\,Vn} = \mathbf{In\,z} \qquad (6.15)$$

where **In** is now a matrix with as many columns as necessary, and **z** a convenient vector for interpretation purposes. Let us work an example.

Example 6.7 *For the circuit of Fig. 6.6 which is reproduced here as Fig. 6.13 for quick reference, find The contributions to the node potentials of the individual sources* 1.3 A *and* 2 A.

Solution *To find the individual contributions, we separate the terms of the known side (in the current vector **In**) which are of the form* $1.3a + 2b$, *and write individual columns* $1.3a$ *and* $2b$ *for each source. Notice that the original elements can be interpreted as* $(1)(1.3) + (-1)(2)$, $(0)(1.3) + (1)(2)$, *and* $(0)(1.3)+(0)(2)$. *That is, we write **In** as*

---

[3]For further information the reader may consult [Kiss68] [Moschytz1974].

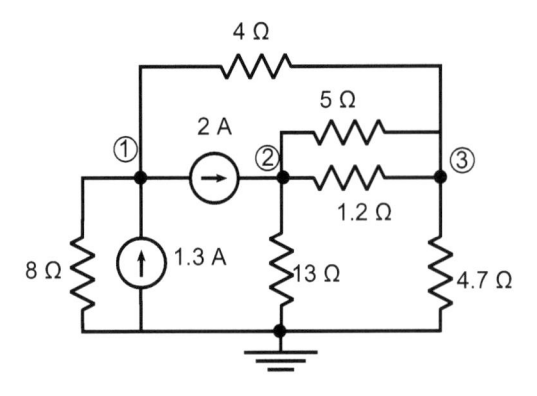

Figure 6.13: Circuit to analyze with nodal method.

$$\mathbf{In} = \begin{bmatrix} 1.3 & -2 \\ 0 & 2 \\ 0 & 0 \end{bmatrix}$$

Using now matrix $\mathbf{Y_n}$ generated before (see eq. 6.11), we have

$$\mathbf{Vn} = \mathbf{Yn}^{-1}\mathbf{In} = \begin{bmatrix} 5.048 & -4.478 \\ 2.137 & 3.730 \\ 2.372 & 1.283 \end{bmatrix} \qquad (6.16)$$

The first column shows the nodal potentials due to the 1.3 A source, while the second column shows those generated by the 2 A source. If you also want the total, you can multiply this result by the numerical vector

$$\mathbf{z} = \begin{bmatrix} 1 \\ 1 \end{bmatrix}$$

6.3.3 Non numerical sources and superposition

For symbolic sources, using 1 as a value; for sources of the form $Kf(t)$, use K. This was explained in Chapter 4, and is still valid. When interpreting, we multiply the result by the symbole used for the source, or by $f(t)$ in the second case. For several sources, apply superposition, using a matrix for the known side instead of a vector, and interpret result using superposition.

Example 6.8 *For the circuit Fig. 6.13 find:*

1. *The node potentials if the 1.3 A and the 2 A sources are instead 1.3 cos(200t) A and 2 sin(120t) A*

2. *The node potentials if these sources are instead* IA *and* IB, *respectively.*

6.4. PROGRAMMING NODAL EQUATIONS I

Solutions:

1) For the sources 1.3 cos(200t) A *and* 2 sin(120t) A *we can use the same matrix* **In** *of example 6.13 and obtain the result (6.16) again. However, now the result should be interpreted as*

$$V_1 = 5.048 \cos(200t) - 4.478 \sin(120t) \text{V}$$
$$V_2 = 2.137 \cos(200t) + 3.730 \sin(120t) \text{V}$$
$$V_3 = 2.372 \cos(200t) + 1.283 \sin(120t) \text{ V}$$

This interpretation follows from the fact that now **z** *is interpreted as*

$$\mathbf{z} = \begin{bmatrix} \cos(200\,t) \\ \sin(200\,t) \end{bmatrix}$$

2) In this case, we use unit values for the sources, and simply interpret the results properly. Set

$$\mathbf{In} = \begin{bmatrix} 1 & -1 \\ 0 & 1 \\ 0 & 0 \end{bmatrix}$$

and apply it to `simult(yn,in)` *or multiply, to solve for*

$$\mathbf{Vn} = \mathbf{Yn}^{-1} \mathbf{In} = \begin{bmatrix} 3.883 & -2.239 \\ 1.644 & 1.865 \\ 1.825 & 0.6415 \end{bmatrix}$$

Therefore,

$$V_1 = 3.883 \,\text{IA} - 2.239 \,\text{IB}$$
$$V_2 = 1.644 \,\text{IA} + 1.865 \,\text{IB}$$
$$V_3 = 1.825 \,\text{IA} + 0.6415 \,\text{IB}$$

Observe that, because of the proportionality property, we could have got the coefficients from the previous item by dividing the first term coefficient by 1.3 and the second one by 2.

Again, the last interpretation follows because now

$$\mathbf{z} = \begin{bmatrix} \text{IA} \\ \text{IB} \end{bmatrix}$$

6.4 Programming nodal equations I

Since our calculator is programmable, let us take advantage to generate the matrices **Yn** and **In** and solve for the equations. The basic algorithm for circuits with characteristics mentioned so far, that is, with resistances, current sources and voltage controlled current sources, is described in the following steps using indefinite admittance matrix. ***Ground is denoted as node "N+1" for easy row and column deletion.***

1. **Number of nodes** Specify number of nodes N, excluding ground. Denote ground as N+1.
2. **Initialization** Initialize matrix $\mathbf{Yn} = 0$ of order (N+1)x(N+1), and vector $\mathbf{IN} = 0$ of order (N+1)x1
3. **Resistance Subroutine** For each resistance of value R ($\neq 0, \infty$) connected between nodes j and k do:
    1. $Yn[j,j] = Yn[j,j] + 1/R$
    2. $Yn[k,k] = Yn[j,j] + 1/R$
    3. $Yn[j,k] = Yn[j,k] - 1/R$
    4. $Yn[k,j] = Yn[k,j] - 1/R$
4. **Voltage Controlled Current Source Subroutine** If there are VCCS's, then, for each source with transconductance g going from node j to node k, and controlled by $(V_p - V_q)$ do:
    1. $Y[j,p] = Y[j,p] + g$
    2. $Y[j,q] = Y[j,q] - g$
    3. $Y[k,p] = Y[k,p] - g$
    4. $Y[k,q] = Y[k,q] + g$
5. **Current source Subroutine** For each current source of value I ($\neq 0, \infty$) going from node j to node k do:
    1. $In[j,1] = In[j,1] - I$
    2. $In[k,1] = In[k,1] + I$
6. **Create definite Yn:** Delete both row and column (N+1) of $\mathbf{Yn}$ and row (N+1) of $\mathbf{In}$
7. **Solve for Vn:** If $|\mathbf{Yn}| \neq 0$ then $\mathbf{Vn} = \mathbf{Yn}^{-1}\mathbf{In}$

In simulators like SPICE, ground is denoted as 0. If you prefer to use such notation, then you should check for ground before each update of elements in the matrices.

Preparing the inputs

We could write our programs in a completely interactive way. I find this rather cumbersome and prefer to describe the circuit before calling the program. The circuit is described using matrices. The number of nodes is **n** excluding ground. Ground is denoted as **n+1** in data.

1. An $m \times 3$ matrix $\mathbf{R}$ for the m resistances in the circuit. If $m = 0$, then enter 0 for this parameter. Each row k consists of three items:
    - Element $(k, 1)$ is a non-zero and non-infinite resistance value,

6.4. PROGRAMMING NODAL EQUATIONS I

- Elements $(k, 2)$ and $(k, 3)$ are the two nodes to which it is connected.

2. A $p \times 3$ matrix **Is** for the p current sources in the circuit. If `Is=0`, then **I**n and **V**n will be 0. Such situation is of interest when we are interested in **Y**n for other uses. Each row k of **Is** consists of three items:

    - Element `is[k,1]` is the finite current source value. It may be 0.
    - Element `is[k,2]` is the node from where the current is leaving,
    - Element `is[k,3]` the node to which it is entering.

3. Input `gs`. This value is 0 if the circuit contains no VCCS. Otherwise, it is a $q \times 5$ matrix. In the k-th row,

    - Element `gs[k,1]` is a finite transconductance g value.
    - Element `gs[k,2]` is the node from where the current is leaving,
    - Element `gs[k,3]` the node to which the current is entering.
    - Element `gs[k,4]` is the node of the positive reference for the controlling voltage, and
    - Element `gs[k,5]` is the node of the negative reference for the controlling voltage.

Fig. 6.14 illustrates how the rows are structured for each of the matrices.

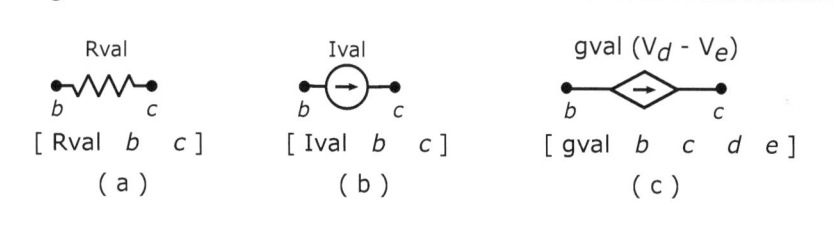

```
        Rval                    Ival                gval (Vd - Ve)
       ┄w┄                     ┄(→)┄                  ┄◇┄
      b     c                  b     c               b     c
   [ Rval  b  c ]           [ Ival  b  c ]        [ gval  b  c  d  e ]
       ( a )                   ( b )                   ( c )
```

Figure 6.14: Inputs structure for program `nodal1`

Examples of descriptions: To illustrate the input circuit description, let us look at some previous examples.

A) For the circuit in Fig. 6.1 on page 97 of example 6.1:

$$n = 3 \quad r = \begin{bmatrix} 8 & 1 & 4 \\ 4 & 1 & 3 \\ 13 & 2 & 4 \\ 1.2 & 2 & 3 \\ 5 & 2 & 3 \\ 4.7 & 3 & 4 \end{bmatrix} \quad \text{is} = \begin{bmatrix} 1.3 & 4 & 1 \\ 2 & 1 & 2 \end{bmatrix} \quad gs = 0$$

B) For the circuit in Fig. 6.10 on page 102 of example 6.3:

$$n = 5 \quad r = \begin{bmatrix} 2000 & 1 & 2 \\ 4000 & 1 & 3 \\ 6000 & 1 & 6 \\ 3.2\text{E}3 + 4.65\text{E}3 & 2 & 4 \\ 2.5\text{E}3 & 4 & 3 \\ 8000 & 3 & 5 \\ 1\text{E}3 & 4 & 5 \\ 6000 & 5 & 6 \\ 2000 & 4 & 6 \end{bmatrix} \quad \text{is} = \begin{bmatrix} 1\text{E}^-3 & 6 & 2 \\ 2\text{E}^-3 & 2 & 3 \end{bmatrix} \quad \text{gs} = 0$$

Observe that in the first column we can specify an equivalent resistance showing the formula for the calculation. Also, I mixed scientific and normal notations to make clear that you are not bound to one type of description.

C) Finally, for the circuit in Fig. 6.11 on page 105 of example 6.4:

$$n = 4 \quad rs = \begin{bmatrix} 6\text{E}3 & 1 & 5 \\ 1.6\text{E}3 & 1 & 2 \\ 4\text{E}3 & 2 & 5 \\ 32000 & 1 & 4 \\ 12\text{E}3 & 2 & 3 \\ 22\text{E}3 & 3 & 5 \\ 1000 & 3 & 4 \\ 120 & 4 & 5 \end{bmatrix}$$

$$\text{is} = \begin{bmatrix} 4\text{E}^-3 & 5 & 1 \end{bmatrix} \quad \text{gs} = \begin{bmatrix} 50/1.6\text{E}3 & 3 & 2 & 1 & 2 \\ 0.015 & 5 & 4 & 5 & 3 \end{bmatrix}$$

Programming the calculator

Fig. 6.15 on the next page shows a program in the TI-nspire cx cas developed using the above pseudo code. It takes as parameters the number of nodes n, excluding ground, and the matrices described in the previous section. In practice, the circuit might be void without any element, but I think you are not looking at this possibility.

The outputs of this program are the nodal admittance matrix yn, the nodal current vector in and the nodal potential vector vn. If there are no current sources, that is, is = 0, then in and vn are 0. You might be interested in these cases when you only need the yn. The program can be easily modified (one line) if you also want the indefinite admittance matrix.

For easy understanding, lines are commented on the right column; these comments are not part of the program but you may include them if you prefer

6.4. PROGRAMMING NODAL EQUATIONS I

```
1    nodal1(n,r,is,gs)
2    Pgrm
3    Local a,b,c,d,e,t
4    newMat(n+1,n+1)→ yn              Initialize the indefinite matrix
5    For t,1,rowDim(r)                Start Yn construction with resistances
6    r[t,1]→a                         a= value of R;
7    r[t,2]→b:r[t,3]→c                R connected to nodes b,c
8    yn[b,b]+1/a→yn[b,b]              Updating $Y_{bb}$
9    yn[c,c]+1/a→yn[c,c]              Update $Y_{cc}$
10   yn[b,c]-1/a→yn[b,c]              Updating $Y_{bc}$ and
11   yn[c,b]-1/a→yn[c,b]              Updating $Y_{cb}$
12   EndFor                           End Resistance subroutine
13   If gs/=0 Then                    Continue Yn when there are VCCS's
14   For t,1,rowDim(gs)               Start introducing dependent sources
15   gs[t,1]→a                        For simpler notation, a=gm;
16   gs[t,2]→b:  gs[t,3]→c            leaving node b, going to node c
17   gs[t,4]→d:  gs[t,5]→e            controlled by Vd-Ve
18   yn[b,d]+a→yn[b,d]                Updating $Y_{bd}$
19   yn[b,e]-a→yn[b,e]                Updating $Y_{be}$
20   yn[c,d]-a→yn[b,d]                Updating $Y_{cd}$
21   yn[c,e]+a→yn[b,e]                Updating $Y_{ce}$
22   EndFor                           End reading gs
23   EndIf                            Finish VCCS subroutine
24   subMat(yn,1,1,n,n)→ yn           Definite admittance matrix
25                                    Blank line for easy reading
26   newMat(n+1,1)→ in                Initializing In
27   If is/=0 Then                    If ind. sources present
28   For t,1,rowDim(is)               Construct In
29   is[t,1]→a                        a=I;
30   is[t,2]→b:is[t,3]→c              from node b to node c
31   in[b,1]-a→in[b,1]                Updating $In_b$
32   in[c,1]+a→in[c,1]                Updating $In_c$
33   EndFor                           End Independent currents subroutine
34   EndIf                            End is/=0 case
35                                    Blank line for easy reading
36   subMat(in,1,1,n,1)→ in           Definite in
37   If det(yn)/=0 Then               If circuit is valid
38   yn^-1*in→ vn                     Solving for nodal voltage vector vn
39   Else                             Otherwise
40   Disp ''Description not valid''   Message for invalid description
41   EndIf                            End vn calculation
42   EndPgrm                          Exit Program
```

Figure 6.15: Example of a program for nodal analysis (resistances and current sources) using TI-nspire cx.

to do so. Similarly, lines are numbered for reference. Except for the case of checking if the circuit description is valid, I have not attempted any extra feature like data checking or other advisable programming precautions for a general case. I wrote the program originally for personal use and I assume that at least I will not provide wrong data.

The reader is encouraged to test the program for the different data. You should also be aware that the program presented here is just an example of a possible listing. The name of the program is automatically included in the variable list displayed with `vas`; it can also be opened from other documents if you store it in public library.

This program does not consider circuits with voltage sources or current dependent sources whose controlling current is not voltage dependent. However, it can be used as a sub routine for a more general situation where these elements are considered.

6.5 Circuits with voltage sources

As stated, nodal analysis is based on using node potentials as voltage variables and writing down the set of current equations for the nodes. Now, a node m may have other elements to it, as illustrated next in Fig. 6.16.

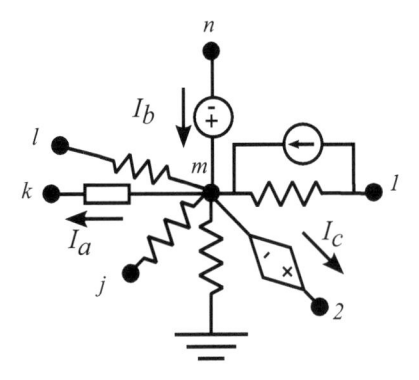

Figure 6.16: Voltage controlled current source and nodal coefficients.

These additional elements such as voltage sources, ideal operational amplifiers, and others, introduce unknown currents which are not voltage dependent. In the figure, these currents are exemplified by I_a, I_b, and I_c. The current equation may now be written as

$$\overbrace{Y_{m1} V_1 + \ldots + Y_{mn} V_n}^{\text{From resistances and VCCS}} + \overbrace{I_a - I_b + I_c}^{\text{non-voltage controlled}} = In_m \qquad (6.17)$$

Observe that the component from resistances and VCCS has already been presented in the previous section and can be written by inspection. The unknown currents are now added, by inspection, to the left hand side (with + sign for currents leaving, - sign for currents entering the node). For each additional current, we need another equation. Common cases for these additional equations are those shown in Table 6.1.

Two approaches can be taken to deal with this type of equations. The first one, traditionally followed in textbooks, is to reduce the number of equations so as to eliminate the presence of these unknown currents. This can be done

6.5. CIRCUITS WITH VOLTAGE SOURCES

either by source transformation as explained in subsection 6.2.6, or else by working with the equations so that no current or only a selected subset of currents appears. In my opinion (*and that's my opinion*), one mistake that textbooks make is not saying explicitly that the procedure is skipping some steps in the system equation solving process. I think this is a mistake because many students find it quite difficult then to solve for those currents when they are needed!

The second and general approach, is the so called Modified Nodal Analysis (MNA)[4], which consists in including all unknowns in the system, as well as the additional equations needed.

Let us start with this general method since the other ones are just a modification of it. Besides, we are using a good calculator! No need to fear having many equations.

Table 6.1: Elements whose current is voltage independent

Element	Symbol	Equation
Grounded Voltage source		$V_m = E$ or take V_m known
Floating V source		$V_m - V_p = E$
V controlled V source		$V_m - V_p = k(V_a - V_b)$
Ideal OA[1]		$Vm - Vp = 0$
C controlled V source[2]		$V_m - V_p - rI_x = 0$
C controlled C source		See note 3

1. Output current shown to illustrate new variable
2. New variable is the source current. V independent I_x was defined elsewhere.
3. You include αI_x in equations of nodes m and p.

---

[4]It receives this name because the first approach was historically the preferred one due to the number of equations that must be solved, among other reasons

6.6 Modified Nodal Analysis (MNA)

For a system of equations to have a unique solution, the number of unknowns must equal the number of equations. Hence,

> For each unknown current present in the circuit, we must have either another equation or a constraint on the set of variables.

From this statement, we set of node equations by inspection for all the nodes, and then we add the additional equations or work the constraints, like those in Table 6.1.

Let us work two examples. The first example is a DC circuit model for a bipolar transistor amplifier.

Example 6.9 *The circuit in Fig. 6.17 is the DC model of a transistor circuit, including parasitics* ICE0 *and* Ro. *For a nodal analysis of the circuit, the unknown currents that must be included are* I_x *and* I_B. *The current controlled source cannot be converted to a voltage controlled one. Hence, the total number of unknowns is 6, assuming for the moment that we do not take* V_1 *as a known variable. Using the values shown in the figure, to simplify writing and keystrokes, we use* $k\Omega$ *units for resistances, and currents in* mA *units (10* μA = .001 mA*)*

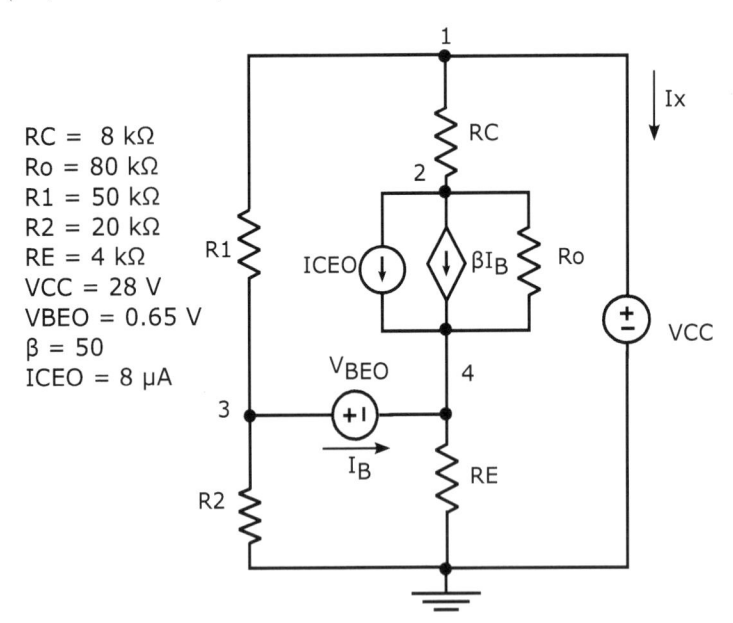

Figure 6.17: A first example for MNA.

Now, first write the individual equations at the nodes, and then the additional constrain equations required to complete the number of necessary

6.6. MODIFIED NODAL ANALYSIS (MNA)

equations. The coefficients for the node potentials follow the same rules as before. For simplicity, I am omitting the terms with coefficient 0, although they should be considered in the calculator.

$$\text{Node 1:} \quad \left(\frac{1}{8} + \frac{1}{50}\right) V_1 - \frac{1}{8} V_2 - \frac{1}{50} V_3 + I_x = 0$$

$$\text{Node 2:} \quad -\frac{1}{8} V_1 + \left(\frac{1}{8} + \frac{1}{80}\right) V_2 - \frac{1}{80} V_4 + 50 I_B = -.008$$

$$\text{Node 3:} \quad -\frac{1}{50} V_1 + \left(\frac{1}{50} + \frac{1}{20}\right) V_3 + I_B = 0$$

$$\text{Node 4:} \quad -\frac{1}{80} V_2 + \left(\frac{1}{80} + \frac{1}{4}\right) V_4 - 51 I_B = .008$$

We have four equations with six variables. Now we add two equations introduced by the voltage sources:

$$V1 = 28$$

and

$$V_3 - V_4 = 0.65$$

The six equations may be incorporated in matrix form as $\mathbf{Ymn}\, \mathbf{X} = \mathbf{Imn}$, – "mn" is for "modified nodal" – as

$$\begin{bmatrix}
 & V_1 & V_2 & V_3 & V_4 & I_x & I_B \\
 & \frac{1}{8}+\frac{1}{50} & -\frac{1}{8} & -\frac{1}{50} & 0 & 1 & 0 \\
\text{Node 2} & -\frac{1}{8} & \frac{1}{8}+\frac{1}{80} & 0 & -\frac{1}{80} & 0 & 50 \\
\text{Node 3} & -\frac{1}{50} & 0 & \frac{1}{50}+\frac{1}{20} & 0 & 0 & 1 \\
\text{Node 4} & 0 & -\frac{1}{80} & 0 & \frac{1}{80}+\frac{1}{4} & 0 & -51 \\
\hline
\text{28 V src} & 1 & 0 & 0 & 0 & 0 & 0 \\
\text{0.65 V src} & 0 & 0 & 1 & -1 & 0 & 0
\end{bmatrix} \times$$

$$\begin{bmatrix} V_1 \\ V_2 \\ V_3 \\ V_4 \\ I_x \\ I_B \end{bmatrix} = \begin{matrix} \text{Node 1} \\ \text{Node 2} \\ \text{Node 3} \\ \text{Node 2} \\ \text{28 V src} \\ \text{0.65 V src} \end{matrix} \begin{bmatrix} 0 \\ -0.008 \\ 0 \\ 0.008 \\ 28 \\ 0.65 \end{bmatrix}$$

The lines showing partition have been introduced to distinguish submatrices in a later explanation. They are not part of the equations, so don't try to enter them in the calculator!!.

With `simult(ymn,imn)` we get the following result, shown here for convenience, in transposed form.

$$\mathbf{X}^T = \begin{bmatrix} V_1 & V_2 & V_3 & V_4 & Ix & I_B \\ 28 & 14.627 & 7.546 & 6.896 & -2.101 & 0.0318 \end{bmatrix}$$

From this result we read $V_1 = 28$ V, $V_2 = 14.627$ V, $V_3 = 7.546$ V, $V_4 = 6.896$ V, $Ix = -2.101$ mA, and $I_B = 0.032$ mA $= 31.8$ μA.

Notice that in the partition of **Ymn Imn** in the previous example, the upper left submatrix of the coefficients as well as the upper subvector in the known side can be written by inspection following the rules already used in subsection 6.2.2. The other columns are really not difficult to write by inspection, but if you have problems with them, first write the second part of the equations separately or else try to visualize them.

Let us work another example.

Example 6.10 *Let us write now the MNA equations for the circuit of Fig. 6.18. This is the circuit used in example 6.6.*

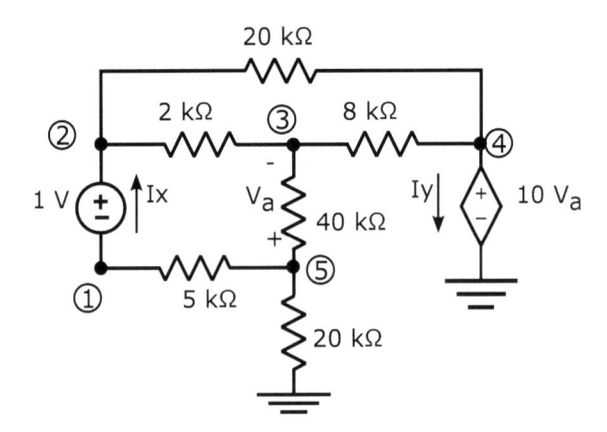

Figure 6.18: Another Example for Modified Nodal Analisis.

Proceeding as we did before, the matrix **Ymn** *and vector* **Imn** *are given below. Again, notice the matrix subdivision shown here for later explanation. Resistance values are again in kΩ units so the currents are in mA units.*

6.6. MODIFIED NODAL ANALYSIS (MNA)

$$\mathbf{Ymn} = \begin{array}{c} \\ Node\ 1 \\ Node\ 2 \\ Node\ 3 \\ Node\ 4 \\ Node\ 5 \\ --- \\ 1V \\ VVCS \end{array} \left[\begin{array}{cccccccc} V_1 & V_2 & V_3 & V_4 & \cdots & V_5 & I_x & I_y \\ \frac{1}{5} & 0 & 0 & 0 & \cdots & -\frac{1}{5} & 1 & 0 \\ 0 & \frac{1}{2}+\frac{1}{20} & -\frac{1}{2} & -\frac{1}{20} & \cdots & 0 & -1 & 0 \\ 0 & -\frac{1}{2} & \frac{1}{40}+\frac{1}{2}+\frac{1}{8} & -\frac{1}{8} & \cdots & -\frac{1}{40} & 0 & 0 \\ 0 & -\frac{1}{20} & -\frac{1}{8} & \frac{1}{20}+\frac{1}{8} & \cdots & 0 & 0 & 1 \\ -\frac{1}{5} & 0 & -\frac{1}{40} & 0 & \cdots & \frac{1}{40}+\frac{1}{5}+\frac{1}{20} & 0 & 0 \\ --- & --- & --- & --- & \cdots & --- & --- & --- \\ -1 & 1 & 0 & 0 & \cdots & 0 & 0 & 0 \\ 0 & 0 & 10 & 1 & \cdots & -10 & 0 & 0 \end{array} \right]$$

$$\mathbf{Imn} = \begin{bmatrix} 0 \\ 0 \\ 0 \\ 0 \\ --- \\ 1 \\ 0 \end{bmatrix}$$

With simult(ymn,imn) we get the following result, shown here for convenience, in transposed form.

$$\mathbf{X}^T = \begin{array}{cccccccc} & V_1 & V_2 & V_3 & V_4 & V_5 & Ix & Iy \\ \left[\right. & -2.801 & -1.801 & -1.960 & -2.548 & -2.215 & 0.118 & -0.111 \left. \right] \end{array}$$

The first five entries are the respective node potentials, and the last two the currents in mA units.

Comparing with example 6.6, we see that potentials V_2, V_3, and V_5 are equal in both cases. Moreover, the two equations for the voltage sources mentioned in that example are included here, as well as the equations for nodes 1 and 4. Thus, this method is more general and we could consider the one in example 6.6 as a system that results after algebraic work on the complete equations. I invite the reader to verify this statement.

Working MNA with known potentials

In example 6.9, we have an independent voltage source of 28 V which is grounded. Hence, $V_1 = 28$ V, as indicated in the constrain equation for the 28 V src. Although the way we proceeded is mathematically correct and also the way that it will result in the programs depicted in the next sections, it is quite annoying to write an additional equation for it, with so many 0's in addition.

An alternate and most common procedure, is to take the potential $V_1 = 28$ already as a known value, eliminating the corresponding equation. This requires "to pass" the first column to the known side with a changed sign and including the value of the potential. The equations would look now as shown next:

$$\begin{array}{c} \\ \text{Node 1} \\ \text{Node 2} \\ \text{Node 3} \\ \text{Node 4} \\ --- \\ 0.65 \text{ V sc} \end{array} \left[\begin{array}{ccc|cc} V_2 & V_3 & V_4 & I_x & I_B \\ -\frac{1}{8} & -\frac{1}{50} & 0 & 1 & 0 \\ \frac{1}{8}+\frac{1}{80} & 0 & -\frac{1}{80} & 0 & 50 \\ 0 & \frac{1}{50}+\frac{1}{20} & 0 & 0 & 1 \\ -\frac{1}{80} & 0 & \frac{1}{80}+\frac{1}{4} & 0 & -51 \\ --- & --- & --- & --- & --- \\ 0 & 1 & -1 & 0 & 0 \end{array} \right] \begin{bmatrix} V_2 \\ V_3 \\ V_4 \\ I_x \\ I_B \end{bmatrix} = $$

$$\begin{array}{c} \text{Node 1} \\ \text{Node 2} \\ \text{Node 3} \\ \text{Node 4} \\ \\ 0.65 \text{ V sc} \end{array} \begin{bmatrix} 0 - \left(\frac{1}{8}+\frac{1}{50}\right)(28) \\ -0.008 + \frac{28}{8} \\ 0 + \frac{28}{50} \\ 0.008 \\ --- \\ 0.65 \end{bmatrix}$$

The reader can verify that, ignoring the first row, the equations obtained correspond to a shifting and source transformation of the VCC source. The complete set adds the equation for node 1 so the current Ix can be obtained.

An advantage of this procedure is that we can already visualize the actual known side and apply superposition faster.

6.6. MODIFIED NODAL ANALYSIS (MNA)

Other cases of known potentials arise for paths of independent sources including ground, as illustrated in Fig. 6.19. Again, "passing" the known potentials to the right side is akin to successive applications of the shifting and conversion theorem, keeping the equations for the currents of the sources.

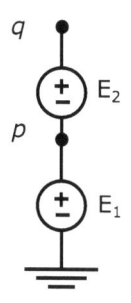

Figure 6.19: Known potentials arising from paths of independent sources. $V_p = E_1$, $V_q = E_1 + E_2$

6.6.1 Programming Modified Nodal Analysis

From examples 6.9 and 6.10, you can verify that **Ymn** and **Imn** may be partitioned as

$$\left[\begin{array}{c|c} \mathbf{Yn} & \mathbf{Ix} \\ \hline \mathbf{VEQ} & \mathbf{0} \end{array}\right] \quad \text{and} \quad \left[\begin{array}{c} \mathbf{In} \\ \hline \mathbf{VsVal} \end{array}\right] \tag{6.18}$$

respectively. Here, **Yn** and **In** are the nodal admittance matrix and nodal current vector already discussed in previous sections. The subroutines for these two submatrices can be found in § 6.4 on page 111, illustrated by the program in Fig. 6.15.

On the other hand, if there are M voltage sources with their respective unknown currents, this set generates submatrices **Ix** of order NxM, **VEQ** of order MxN, and the subvector **VsVal** of order M. N is the number of nodes excluding ground. These submatrices and vector arise from the M constraint equations added to the nodal equations.

Finally, **0** is a zero-matrix of order MxM. If the circuit contains current controlled voltage sources (CCVS) where the controlling current is voltage independent, then this will not be a zero matrix. This case is not considered here.

Remark: We will assume that current controlled current sources (CCCS) in which the current of control cannot be related to node potentials have the controlling current in a voltage source. This is illustrated in example 6.9 with current IB. Any other case can be worked by dealing with a manual edition of the matrices.

Since we already have subroutines for **Yn** and **In**, the pseudocode below works building matrices **Ymn** and **Imn** by matrix composition. To facilitate our task, we adopt the following

Convention: By definition, we assume that in the voltage sources the current is described by the passive convention of power, that is, it goes from the positive terminal to the negative terminal.

In other words, *for programming purposes, we will assume the current in the sources going from the + terminal to the negative terminal.* This means that if the current yields a negative value, the source is generating power and actual direction is leaving the positive terminal. This is a usual convention.

Circuit Description

Our pseudocode works with the description of the circuit using the number of nodes and a set of matrices as described below. The number of nodes N, as well as matrices **R**,**Is**, and **Gs** have been described before, but are repeated below for convenience. What we add for our analysis are the matrices that describe the way the voltage sources are connected. The set

Number of nodes The number of nodes N excluding ground

Matrix R for resistances For m resistances, an mx3 matrix, where each row describes a resistance connection using the same convention illustrated in Fig. 6.14. The matrix is 0 if there are no resistances.

Matrix Is for independent current sources For p independent current sources, a px3 matrix, where each row describes a source and its connections using the same convention illustrated in Fig. 6.14. The matrix is 0 if there are no independent current sources.

Matrix Gs for voltage dependent current sources For q VCCS, a qx5 matrix, where each row describes the dependent source and its connections using the same convention illustrated in Fig. 6.14. The matrix is 0 if there are no voltage controlled current sources.

Matrix Vs for independent voltage sources For s independent voltage sources, <u>whose currents do not control CCCS</u>, this is an sx3 matrix, where each row describes a source and its connections using the convention illustrated in Fig. 6.20. The matrix is 0 if there are no independent voltage sources.

Matrix Ks for voltage dependent voltage sources For t VCVS, a tx5 matrix, where each row describes a source and its connections using the convention illustrated in Fig. 6.20(b). The matrix is 0 if there are no dependent voltage sources.

6.6. MODIFIED NODAL ANALYSIS (MNA)

Matrix CCs for current dependent current sources For k CCCS, each one being controlled by a current in a voltage source, define a a kx6 matrix, where each row describes a voltage source and its connections, and the CCC source, using the convention illustrated in Fig. 6.20(c). The matrix is 0 if there are no dependent current controlled sources.[5]

We can work other situations by hand editing the matrices. Let us now illustrate the circuit description with examples.

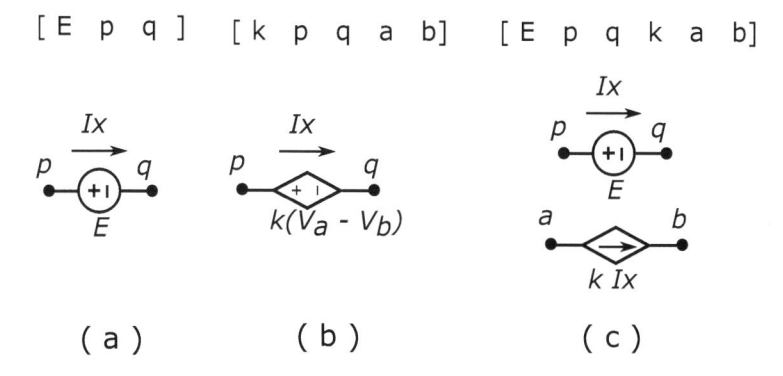

Figure 6.20: Matrix description of voltage sources for Modified Nodal Analisis.

Example 6.11 *The circuit of example 6.9, shown in 6.17 on page 118 is described by the following matrices:*

$$\mathbf{R} = \begin{bmatrix} 50E3 & 1 & 3 \\ 20E3 & 3 & 5 \\ 4E3 & 4 & 5 \\ 8E3 & 1 & 2 \\ 80E3 & 2 & 4 \end{bmatrix} \quad \mathbf{Is} = [\; 8E\;(\text{-})\;6 \quad 2 \quad 4 \;] \quad \mathbf{Gs} = 0$$

$\mathbf{Vs} = [\; 28 \quad 1 \quad 5 \;] \quad \mathbf{Ks} = 0 \quad \mathbf{CCs} = [\; 0.65 \quad 3 \quad 4 \quad 50 \quad 2 \quad 4 \;]$

On the other hand, the circuit of example 6.10, shown in 6.18 on page 120 is described by the following matrices:

---

[5]We assume a non-zero voltage source here. SPICE, the popular simulater, uses 0 V sources for these controlled sources

$$\mathbf{R} = \begin{bmatrix} 5E3 & 1 & 5 \\ 2E3 & 2 & 3 \\ 20E3 & 2 & 4 \\ 8E3 & 3 & 4 \\ 40E3 & 3 & 5 \\ 20E3 & 5 & 6 \end{bmatrix} \quad \mathbf{Is} = 0 \quad \mathbf{Gs} = 0$$

$$\mathbf{Vs} = [\,1\ \ 2\ \ 1\,] \quad \mathbf{Ks} = [\,10\ \ 4\ \ 6\ \ 5\ \ 3\,] \quad \mathbf{CCs} = 0$$

Program pseudocode for mna(n,r,is,gs,vs,ks,ccs)

One pseudocode, assuming the inputs described before, is the following:

Step 1 Generate submatrices **Yn** and **In** calling the program nodal1(n,r,is,gs) as subroutine.

Check for need of MNA analysis:

Step 2 IF (**Vs**=0 AND **Ks**=0 AND **CCS**=0) THEN
        Display message "No voltage sources"
        STOP
        ENDIF

Initialize submatrices and pointer index:

Step 3 Determine value M, for the size of matrices

Step 4 **VEQ** = 0 of order Mx(N+1), **Ix** = 0 of order (N+1)x M and vector **Vsval** = 0 of order Mx1

Step 5 SourceIndex=0 for indexing purposes

Enter independent voltage sources:

Step 6 IF **Vs** $\neq$ 0 THEN
    FOR $h = 1$ to rowdimension(**Vs**)
      SourceIndex = SourceIndex+1
      E= **Vs**[h,1], p= **Vs**[h,2], and q= **Vs**[h,3]
      VEQ[SourceIndex,p] = 1
      VEQ[SourceIndex,q] = -1
      VsVal[SourceIndex,1] = E
      Ix[p,SourceIndex] = **Ix**[p,SourceIndex] + 1
      Ix[q,SourceIndex] = **Ix**[q,SourceIndex] -1
    ENDFOR

6.6. MODIFIED NODAL ANALYSIS (MNA)

ENDIF

Enter voltage dependent voltage sources:

Step 7 IF **Ks** $\neq$ 0 THEN
FOR h = 1 to rowdimension(**Ks**)
SourceIndex = SourceIndex+1
K= VS[h,1], p= **Ks**[h,2], and q= **Ks**[h,3]
a= **Ks**[h,4] and b= **Ks**[h,5]
VEQ[SourceIndex,p] = **VEQ**[SourceIndex,p] + 1
VEQ[SourceIndex,q] = **VEQ**[SourceIndex,q]-1
VEQ[SourceIndex,a] = **VEQ**[SourceIndex,a] - k
VEQ[SourceIndex,b] = **VEQ**[SourceIndex,b] + k
Ix[p,SourceIndex] = **Ix**[p,SourceIndex] + 1
Ix[q,SourceIndex] = **Ix**[q,SourceIndex] -1
ENDFOR
ENDIF

Enter current dependent current sources:

Step 8 IF **CCs** $\neq$ 0 THEN
FOR h = 1 to rowdimension(**CCs**)
SourceIndex = SourceIndex+1
E= **CCs**[h,1], m= **CCs**[h,2], p = **CCs**[h,3]
K= **CCs**[h,4], a= **CCs**[h,5], b = **CCs**[h,6]
VEQ[SourceIndex,m] = 1
VEQ[SourceIndex,p] = -1
VsVal[SourceIndex,1] = E
Ix[m,SourceIndex] = **Ix**[m,SourceIndex] + 1
Ix[p,SourceIndex] = **Ix**[p,SourceIndex] -1
Ix[a,SourceIndex] = **Ix**[a,SourceIndex] + K
Ix[b,SourceIndex] = **Ix**[b,SourceIndex] - K
ENDFOR
ENDIF

Adjust dimensions of submatrices:

Step 9 Delete (N+1)-th column of **VEQ**, and row (N+1) of **Ix**

Compose Ymn and Imn and solve:

Step 10 Compose **Ymn** and **Imn** of expression (6.18) – See subsection §3.4.6 on page 37 –

11.1 p= Augment(Yn,Ix)

11.2 m= Augment(VEQ,0)

11.3 Ymn = Augment(p;m)

11.4 Imn = Augment(In;VsVal)

Step 11 IF det(**Ymn**)$\neq$ 0 THEN
    Vmn = **Ymn**$^{-1}$ **Imn**
  ELSE
    Display "Matrix **Ymn** ill defined"
  ENDIF

Exercise As an exercise for the reader, I suggest to use the above pseudocode and test it with the examples already presented.

Developing the program for mna

Since the program is rather large, for the sake of easy reading and printing, let me present it in parts. Remember that some devices have not been included, this is an exercise in learning. And besides, the pseudocode and program are still very useful for a lot, if not most, of circuits you may find along the way. In any case, you can always edit by hand the matrices. Or you can always add the devices.

```
1    mna1(n,r,is,gs,vs,ks,ccs)
2    Pgrm
3    Local a,b,p,q,e,k,h,sourceindex
4    nodal1(n,r,is,gs)              Generate Yn and In
5    0 → h                          Determine size of submatrices
6    If vs/=0 Then
7      h+ rowDim(vs) → h            Number of
8    Endif                          independent V sources
9    If ks/=0 Then                  plus
10     h+ rowDim(ks) → h            Number of
11   Endif                          VCV sources
12   If ccs/=0 Then                 plus
13     h+ rowDim(ccs) → h           Number of
14   Endif                          CCS sources
15   If h=0 Then                    If no need of MNA
16     Disp ''No voltage sources'' :stop   leave program
77   Endif
18   newMat(h,n+1)→ veq             Initialize VEQ
19   newMat(n+1,h)→ ix              Initialize Ix
20   newMat(h,1)→ vsval             Initialize VsVal
```

Figure 6.21: First part of the modified nodal analysis program using the given pseudocode.

Fig. 6.21 shows the first part of the program. It covers the generation of **Yn** and **In**, as well as the initialization of submatrices **VEQ**, **Ix** and **VsVal**.

6.6. MODIFIED NODAL ANALYSIS (MNA)

To do it, we need the size of these submatrices. Specifically, the number of rows for **VEQ** and **VsVal**, which is the same as the number of columns for **Ix**. This number is equal to (number of independent voltage sources + number of voltage controlled voltage sources + number of current controlled current sources) and is determined in lines 5 to 14.

Lines 15, 16, and 17 were included just for the sake of completeness.

```
21   0 → sourceindex                              Init. Index
22   If vs/=0 Then                                Introduce Ind. Sources
23   For h,1,rowDim(vs)                           Start
24   sourceindex + 1 → sourceindex                update index
25   vs[h,1]→ E                                   source value E;
26   vs[h,2] → p :vs[h,3]→ q                      terminals +p -q
27   +1→ veq[sourceindex,p]                       Updating Veq
28   -1→ veq[sourceindex,q]
29   E→ vsval[sourceindex,1]                      Update VsVal
30   ix[p,sourceindex]+1→ ix[p,sourceindex]       Updating Ix
31   ix[q,sourceindex]-1→ ix[q,sourceindex]
32   EndFor :Endif                                End independent sources.
33   If ks/=0 Then                                Start dependent VCVS
34   For h,1,rowDim(vs)                           Start
35   sourceindex + 1 → sourceindex                update index
36   vs[h,1]→ K                                   Gain K;
37   vs[h,2] → p:  vs[h,3]→ q                     terminals +p -q
38   ks[h,4]→a:  ks[h,5]→b                        controlled by Va-Vb
39   veq[sourceindex,p]+1→ veq[sourceindex,p]     Updating Veq
40   veq[sourceindex,q]-1→ veq[sourceindex,q]
41   veq[sourceindex,a]-K→ veq[sourceindex,a]
42   veq[sourceindex,b]+K→ veq[sourceindex,b]
43   ix[p,sourceindex]+1→ ix[p,sourceindex]       Updating Ix
44   ix[q,sourceindex]-1→ ix[q,sourceindex]
45   EndFor :Endif                                End VCV sources.
46   If ccs/=0 Then                               Introduce CCS sources
47   For h,1,rowDim(ccs)                          Start
48   sourceindex + 1 → sourceindex                update index
49   vs[h,1]→ E                                   controlling source value E;
50   vs[h,2] → p :vs[h,3]→ q                      terminals +p -q
51   +1→ veq[sourceindex,p]                       Updating Veq
52   -1→ veq[sourceindex,q]
53   E→ vsval[sourceindex,1]                      Update VsVal
54   ix[p,sourceindex]+1→ ix[p,sourceindex]       Updating Ix
55   ix[q,sourceindex]-1→ ix[q,sourceindex]
56   ccs[h,4]→ K                                  Gain K of CCS;
57   ccs[h,5]→a :ccs[h,6]→b                       from a to b
58   ix[a,sourceindex]+K→ ix[a,sourceindex]       Updating Ix
59   ix[b,sourceindex]-K→ ix[b,sourceindex]
60   EndFor :Endif                                End CCS sources.
```

Figure 6.22: Second part of mna program, entering the elements with unknown currents.

The second part of the program, shown in Fig. 6.22, introduces the sources. It is in this section that you may add lines to include those devices not considered. The (local) variable `sourceindex` is used to signal the row for the constraint equation and the column for the unknown current within the respective matrices. Notice that they are placed in the same order in which you have described them in the data matrices.

The third part of this code is shown in Fig. 6.23. These lines compose **Ymn** and **Imn**, and solve the equations.

```
61   subMat(veq,1,1,sourceindex,n)→ veq      adjust veq
62   subMat(ix,1,1,n,sourceindex)→ ix        Definite ix
63   augment(yn,Ix) → p                      Upper subpartition of Ymn
64   newMat(sourceindex,sourceindex) → q     Lower subpartition
65   augment(veq,q)→ q
66   augment(p;q) → Ymn                      Compose Ymn
67   augment(in;vsval) → Imn                 Compose Imn
68   If det(Ymn)/=0 Then                     If circuit is valid
69   Ymn∧ (-) 1 *Imn→ Vn                     Solve equations
70   Else                                    Otherwise
71   Disp ''Description not valid''          Message for invalid description
72   EndIf                                   End vn calculation
73   EndPrgm                                 Exit Program
```

Figure 6.23: Last part of mna program, composing and solving.

Well, if everything went fine and neither you nor I did something wrong, test the program. Remember though that the most important step in the process is to interpret results. In particular, your solution `Vn` includes both node potentials and voltage source currents.

6.7 "Reducing" number of equations

Most textbooks, at least the introductory ones, avoid modified nodal analysis. Again, the reason is simple: they are oriented to hand analysis, so the student should focus in setting up systems with as few equations and variables as possible. This may be achieved by working directly on the equations, so as to reduce variables. But textbooks' authors usually prefer to work with the circuits only and write equations directly. Two situations are considered here. The first one is by completely eliminating unknown currents. The second one consists in leaving one ore more currents in the set.

6.7.1 Using Supernodes

The concept of "supernodes", a term that I also use because it is well established, but don't like[6], is illustrated with Fig. 6.24, where the nodes inside the closed dashed line constitute the supernode.

---
[6] Actually, it is a concept known to circuit and graph theoreticians as "cut-set". But someone in the academic community introduced the term supernode, which has gained its place in basic circuit analysis textbooks.

6.7. "REDUCING" NUMBER OF EQUATIONS

$$\text{(p,q)} \;[\;..\;Y_{pp}\;...\;Y_{qq}...Y_{pj}+Y_{qj}\;..\;] \begin{bmatrix} V_p \\ V_q \\ V_j \end{bmatrix} \; [I_{np}+I_{nq}]$$

(a)

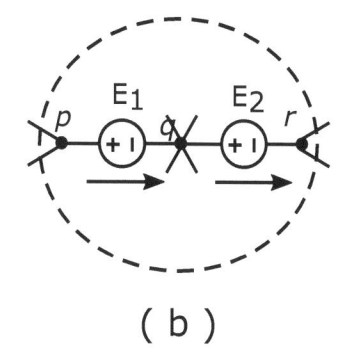

$$\text{(p,q,r)} \;[\;..\;Y_{pp}\;..\;Y_{qq}\;..Y_{rr}\;\;Y_{pj}+Y_{qj}+Y_{rj}\;..\;] \begin{bmatrix} V_p \\ V_q \\ V_j \end{bmatrix} \; [I_{np}+I_{nq}+I_{nr}]$$

(b)

Figure 6.24: Supernode concept and the respective equations. If one node is ground, no equation is taken.

The figure includes the equation assigned to the supernode, *assuming no resistances shared by the nodes within the closed curve.* If this is not the case, substitute Ypp and Yqq by Ypp + Ypq, and Yqq + Yqp.

The method consists in not taking equations at nodes p, q, r,... thereby eliminating the currents, and only using the equation for the supernode. *You cannot use supernodes if the current in the voltage source controls a dependent source!* This remark is illustrated by example 6.9, where the unknown current I_B controls a dependent source.

The reader can demonstrate that, mathematically, the equation of the supernode is equal to the addition of the equations for the nodes inside the closed curve. On the other hand, if one of the nodes is ground, we can eliminate the set of equations for this group because an unknown current will appear only in one equation, which then becomes unnecessary to solve for the rest of the variables.

Let us work examples on writing the equations directly with the knowledge you have gained, saving time and effort while introducing equations in your calculator. Remember, try more exercises from your textbook of from other circuit books.

Example 6.12 *Let us start with the circuit of Fig. 6.18 on page 120 from example 6.10. Node 4 belongs to a grounded voltage source, so no equation is taken at this node. On the other hand, nodes 1 and 2 constitute a supernode created with the 1 V source. With this information, we may write the matrix equation* **Yn Vn** = **In**, *where* **Yn** *and* **In** *are as shown below. Notice that there are five equations, just for the five potentials. For the the equation of supernode, labeled as* SN. (1,2), *apply the principles shown in Fig. 6.24.*

$$\mathbf{Yn} = \begin{array}{c} SN.\ (1,2) \\ N.\ 3 \\ N.\ 5 \\ 1V \\ VV \end{array} \begin{bmatrix} V_1 & V_2 & V_3 & V_4 & V_5 \\ \frac{1}{5} & \frac{1}{2}+\frac{1}{20} & \frac{1}{2} & -\frac{1}{20} & -\frac{1}{5} \\ 0 & -\frac{1}{2} & \frac{1}{40}+\frac{1}{2}+\frac{1}{8} & -\frac{1}{8} & -\frac{1}{40} \\ -\frac{1}{5} & 0 & -\frac{1}{40} & 0 & \frac{1}{40}+\frac{1}{5}+\frac{1}{20} \\ -1 & 1 & 0 & 0 & 0 \\ 0 & 0 & 10 & 1 & -10 \end{bmatrix}$$

$$\mathbf{In} = \begin{bmatrix} SN.(1,2) & N.3 & N.5 & 1V & VV \\ 0 & 0 & 0 & 1 & 0 \end{bmatrix}^T$$

I encourage the reader to solve this matrix equation and verify the the node potentials are the same as before.

Let us next work another example, this time with a known potential. For the known potential, "we pass it" to the known side, as always.

Example 6.13 *This example contains two grounded sources, one is known and the other is a dependent source. The circuit is shown in Fig. 6.25. In this circuit, I have shown explicitly the currents in the voltage sources. The current in the independent source will appear exclusively in the equation of node 4. Hence we can eliminate this equation. On the other hand, the equation of node 3 is substituted by the equation of the VCVS. This can also be easily done.*

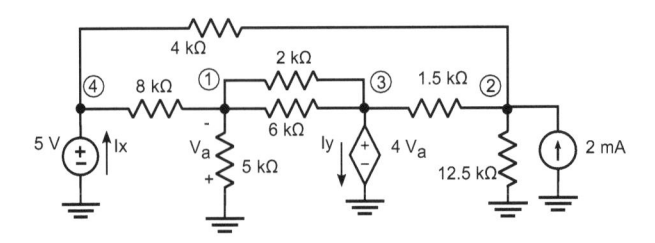

Figure 6.25: Example: circuit with two grounded voltage sources

6.7. "REDUCING" NUMBER OF EQUATIONS

The voltage controlled voltage source $4v_a$ is characterized by the constraint $V_3 = -4V_1$ or, taking into consideration that both potentials V_1 and V_3 are unknowns, by

$$4V_1 + V_3 = 0$$

This equation is to be included in the set, instead of the equation at node 3.

Let us now write the equations **by inspection** considering the following facts:

- The unknown potentials are V_1, V_2 and V_3. Hence the coefficient matrix has three columns

- There are equations for nodes 1 and 2, but not for node 3. These equations follow the normal rules explained before.

- Instead of an equation for node 3, we use the equation for the controlled voltage source

The equation uses kΩ units for resistances, and mA units for current sources.

$$\begin{array}{c} Node1 \\ Node2 \\ VCVS \end{array} \begin{bmatrix} \overset{V_1}{\frac{1}{8}+\frac{1}{5}+\frac{1}{2}+\frac{1}{6}} & \overset{V_2}{0} & \overset{V_3}{-\frac{1}{2}-\frac{1}{6}} \\ 0 & \frac{1}{1.5}+\frac{1}{12.5}+\frac{1}{4} & -\frac{1}{1.5} \\ 4 & 0 & 1 \end{bmatrix} \begin{bmatrix} V_1 \\ V_2 \\ V_3 \end{bmatrix} = \quad (6.19)$$

$$\begin{array}{c} Node1 \\ Node2 \\ VCVS \end{array} \begin{bmatrix} "I_n - 5\,Y_{k4}" \\ \frac{5}{8} \\ 2+\frac{5}{4} \\ 0 \end{bmatrix}$$

I suggest the reader to write the full MNA equations and then apply mathematical procedures that will reduce the system to that obtained in the example. Let us work another example.

Applying superposition

As you might imagine, we can also apply superposition in all this cases. Remember that we can apply it directly by splitting the terms of the known side in columns. Let's look at the next example.

Example 6.14 *Superposition*

Let us now write the previous example looking for the individual source contributions to the node voltages. Each term in the right hand vector of (6.19) is a sum of the form $b(2) + 5a$, where $b = 0$ in the first and third equation. The first term shows the contribution of the 2 mA source to the term, the second one that of the 5 V source. Writing for superposition means the individual terms in columns, writing the equation now in the form

$$\begin{array}{c} Node1 \\ Node2 \\ VCVS \end{array} \begin{bmatrix} \overset{V_1}{\frac{1}{8} + \frac{1}{5} + \frac{1}{2} + \frac{1}{6}} & \overset{V_2}{0} & \overset{V_3}{-\frac{1}{2} - \frac{1}{6}} \\ 0 & \frac{1}{1.5} + \frac{1}{12.5} + \frac{1}{4} & -\frac{1}{1.5} \\ 4 & 0 & 1 \end{bmatrix} \begin{bmatrix} V_1 \\ V_2 \\ V_3 \end{bmatrix} = $$
(6.20)

$$\begin{array}{c} Node1 \\ Node2 \\ VCVS \end{array} \begin{bmatrix} I_n & -5Y_{k4} \\ 0 & \frac{5}{8} \\ 2 & \frac{5}{4} \\ 0 & 0 \end{bmatrix}$$

Once we have the correct settings, we can proceed as necessary, as it was done in example 6.7.

6.7.2 "Hybrid" reduced MNA

Section 6.6 deals with equations when we want all the unknown currents. On the other hand, the previous subsection applies when we don't want any current, except when forced by the circuit itself. Well, in particular for our purposes, we could apply the principles of reduction only in a partial form so that we keep in the system specific unknown currents of interest. In later chapters we will see why these situations are important. Even more, fundamental for an efficient use of the calculator.

Whenever we need one or more specific currents, what we to do is to also take the equations where those currents appear. The following example illustrates this remark.

Example 6.15 *Take again the circuit from Fig. 6.25. If we want, for any reason, the current of the independent voltage source, all we have to do is to include the equation of node 4 in the set. We could also work example 6.13 and solve equation of node 4 separately. But then, what do we want the calculator for?*

6.7. "REDUCING" NUMBER OF EQUATIONS

$$\begin{array}{c} \\ Node1 \\ Node2 \\ Node4 \\ VCVS \end{array} \begin{bmatrix} V_1 & V_2 & V_3 & I_x \\ \frac{1}{8}+\frac{1}{5}+\frac{1}{2}+\frac{1}{6} & 0 & -\frac{1}{2}-\frac{1}{6} & 0 \\ 0 & \frac{1}{1.5}+\frac{1}{12.5}+\frac{1}{4} & -\frac{1}{1.5} & 0 \\ -\frac{1}{8} & -\frac{1}{4} & 0 & -1 \\ 4 & 0 & 1 & 0 \end{bmatrix} \begin{bmatrix} V_1 \\ V_2 \\ V_3 \\ I_x \end{bmatrix} =$$

$$\begin{array}{c} Node1 \\ Node2 \\ Node4 \\ VCVS \end{array} \begin{bmatrix} \text{``}I_n - 5\, Y_{k4}\text{''} \\ \frac{5}{8} \\ 2+\frac{5}{4} \\ -\frac{1}{8}-\frac{1}{4} \\ 0 \end{bmatrix}$$

(6.21)

We could work the problem also with superposition in mind.

As the reader might imagine, it is possible to program simplified nodal analysis. I leave this exercise to the interested reader.

6.7.3 Working with Operational Amplifiers

I close this chapter with comments and examples of nodal analysis for circuits containing ideal operational amplifiers. I could dedicate one chapter to this type of circuits or extend the chapter for several more pages. At this point, however, let us illustrate the main ideas with one circuit. The process I present is oriented toward hand analysis writing the equations directly .... of course!

The reason why I deal with OA separately is that, in a certain way, we work again with a sort of "supernode", but this time applied to the potential nodes, not to the equations! Let's see what I mean next. The principles governing this procedure are the following:

OA output: Unless you are interested on the current at the output of an OA, you do not take an equation at the output node.

OA input nodes potentials: All nodes connected through a chain of OA inputs are assigned the same variable. If one of the terminals is connected to a known voltage – including ground –, it is worked as a known voltage.

Equations at nodes: They follow the same principles as before, except that you should use the same variable, hence the same column, for all nodes connected by OA inputs.

Let us work an example using symbolic values so we can concentrate on the above rules. After that we work a numerical example.

Example 6.16 *The circuit in Fig. 6.26 has two OA's. Write the nodal equations.*

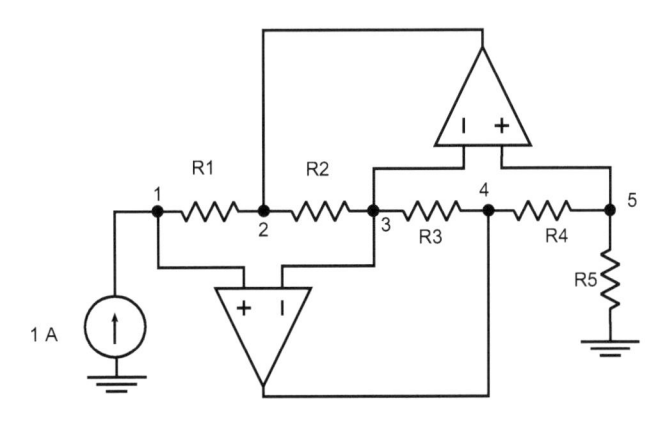

Figure 6.26: A circuit with Operational amplifiers

Nodes 1, 3 and 5 are all connected by OA inputs. Hence we use one variable, or one column, for the three potentials, because $V_1 = V_3 = V_5$. Let us use V_1. For the coefficients, apply the rule of (6.10) on page 97. For easy writing and problem solving, I use the conductance G_j instead of $1/R_j$

$$\begin{array}{c} \\ \text{Node 1} \\ \text{Node 3} \\ \text{Node 5} \end{array} \begin{array}{c} V_1(V_3, V_5) \quad V2 \quad V4 \\ \begin{bmatrix} G_1 & -G_1 & 0 \\ G_2 + G_3 & -G_2 & -G_3 \\ G_4 + G_5 & 0 & -G_4 \end{bmatrix} \end{array} \begin{bmatrix} V_1 \\ V_2 \\ V_4 \end{bmatrix} = \begin{bmatrix} 1 \\ 0 \\ 0 \end{bmatrix}$$

Now, if available, you may work the problem with the symbolic capabilities of the calculator. Using the command entry line –although you can use the templates key –, enter

yn:= [g1, (-)g1,0 ;g2+g3, (-)g2, (-)g3; g4+g5,0,(-)g4]

in := [1,0,0]^T

yn^{-1} × in

to get the result

6.7. "REDUCING" NUMBER OF EQUATIONS

$$V1=V3=V5 \quad \left[\frac{g2 \cdot g4}{g1 \cdot g3 \cdot g5} \right]$$

$$V2 \quad \left[\frac{g2 \cdot g4 - g3 \cdot g5}{g1 \cdot g3 \cdot g5} \right]$$

$$V4 \quad \left[\frac{g2(g4 + g5)}{g1 \cdot g3 \cdot g5} \right]$$

If you want everything in terms or resistances, multiply numerators and denominators by R1 R2 R3 R4 R5.

Now a numeric example.

Example 6.17 *For the two OA circuit shown in Fig. 6.27, find the output voltage at node 2. The circuit has two inputs with known potentials* Va *and* Vb.

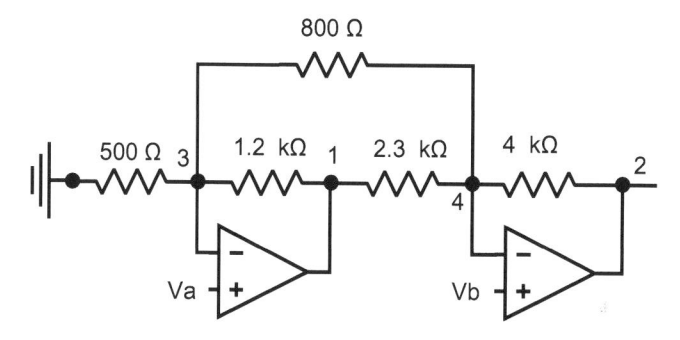

Figure 6.27: A circuit with Operational amplifiers

From the properties of ideal operational amplifiers, V3 = Va *and* V4 = Vb. *Since we are not interested on the output currents of the amplifiers, we take equations only at nodes 3 and 4. Applying the rules that we have seen in the chapter, and scaling resistances to* kΩ *units, we can write the equations as*

$$\begin{array}{c} \text{Node 3} \\ \text{Node 4} \end{array} \begin{bmatrix} V1 & V2 \\ -\frac{1}{1.2} & 0 \\ -\frac{1}{2.3} & -\frac{1}{4} \end{bmatrix} \begin{bmatrix} V_1 \\ V_2 \end{bmatrix} = \begin{bmatrix} \text{Va (=V3)} & \text{Vb (=V4)} \\ -\frac{1}{0.8} - \frac{1}{1.2} - \frac{1}{0.5} & \frac{1}{0.8} \\ \frac{1}{0.8} & -\frac{1}{0.8} - \frac{1}{1.2} - \frac{1}{2.4} \end{bmatrix}$$

which we solve as

$$\begin{bmatrix} V_1 \\ V_2 \end{bmatrix} = \begin{bmatrix} \text{Va} & \text{Vb} \\ 4.9 & -1.5 \\ -13.522 & 12.609 \end{bmatrix}$$

Therefore, V2 = 12.609 Vb -13.522 Va.

6.7.4 Closing remarks

This chapter has included much more material that what is covered in textbooks. One characteristic of nodal analysis, both for the modified one and the traditional case, is that they are relatively easy to program. And also the process is quite direct when entering directly the data into matrices.

You will find that this chapter is longer than the one dedicated to loop analysis. One reason is that, although not impossible, general loop analysis is not so direct to program. Another one is that nodal analysis can be extended beyond those topics explained here so that embedded subcircuits, such as for example the equivalent subcircuit of transistors, operational amplifiers, and others, is relatively easy. I do not deal with this topic, but interested readers may consult [Moschytz1974].

Since writing equations is a very important step in solving circuit problems, I suggest the reader to practice, practice, and practice.

CHAPTER 7

Loop Analysis

Loop analysis is as popular as nodal analysis among students. Perhaps more because it does not involve conductances. But then again, with calculators this should not be a problem.

I present the loop analysis method in two steps. First, following the traditional textbook approach working with circuits containing only resistances, known voltages and voltage dependent current sources. I include a program in mesh analysis In the second stage, I include unknown voltages not dependent on currents.

7.1 Loop currents and loops selection

This section introduces the concepts around loop currents. The reader may skip it and come back if necessary.

A loop is a closed path in a circuit, where no node is touched twice. In planar networks drawn without crossing lines, a mesh is any loop which separates the plane in two half planes, one of which does not contain any element. Most examples in textbooks and also here, use planar examples. But that does not mean we are constrained to use meshes, or limited to planar circuits. A pair of examples will show this.

A *loop current* is a mathematical entity envisioned as a current which is constrained to circulate within a loop, as illustrated in Fig. 7.1(a). If the loop is a mesh, the current is also called *mesh current*. To set up the loop equations, we must select a set of B-N+1 independent loops where B is the number of elements and N the number of nodes in the circuit, including those for elements in series. In planar networks and working with meshes, it suffices to discard one of the meshes to have an independent set. For both planar and non planar circuits, there are several algorithms that allow us to select an independent set.

CHAPTER 7. LOOP ANALYSIS

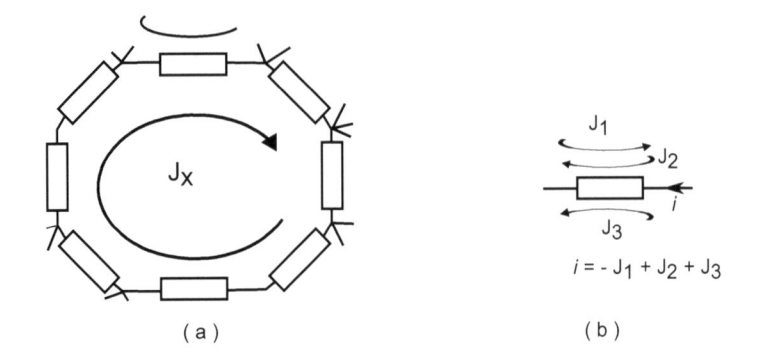

Figure 7.1: (a) Loop Current Concept. (b) Current in element as function of loop currents

Any element must be in at least one loop, but it may belong to two or more loops. The current in the element is associated to the respective loop currents in the form illustrated by Fig. 7.1(b). Namely, the current in the element is the algebraic sum of the loop currents passing through it, with positive sign (+) if the loop current goes in the same direction as the element current, and minus sign (-) if it goes in the opposite direction,

$$i = \sum_{\text{same direction}} J_x - \sum_{\text{Opposite direction}} J_y \qquad (7.1)$$

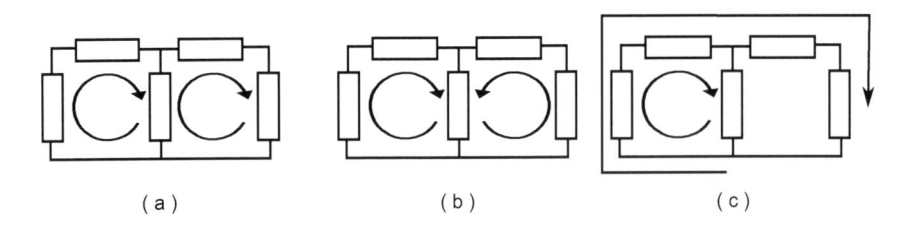

Figure 7.2: Three Mesh circuit: (a) The two internal meshes selected; (b) same meshes with different directions; (b) One internal mesh and the external mesh selected.

In nodal analysis, once the reference node is selected, all equations are defined and potentials have physical meaning. Physical meaning for a loop current is possible only if there is an element in the loop that is not shared with other loops. Moreover, in loop analysis, we must select the set of loops and the direction of the currents, i. e., clockwise or counterclockwise. Fig. 7.2 illustrates this statement for a planar circuit with three meshes. Any two of these form an independent set. Hence, we have three possible independent

7.1. LOOP CURRENTS AND LOOPS SELECTION

sets. Yet, when direction is considered, there are twelve different cases of loop current selection. The figure shows only three of them.

Fig. 7.3 illustrates another popular planar topology with four meshes. Not considering loop directions, there are 16 sets of three independent loops. Including directions, the number becomes 128. The figure shows just two of these possibilities. Case (a) is the popular mesh selection. Notice that (b) includes a loop which is not a mesh and, moreover, there are elements which are shared by all three independent loops.

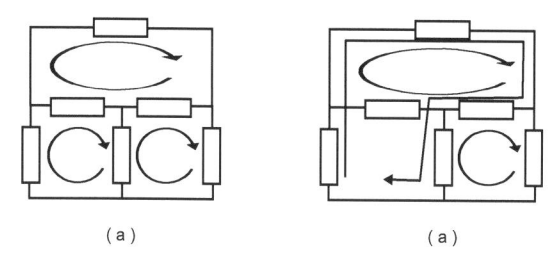

(a)   (a)

Figure 7.3: Another example of independent loops selection.

The basis for direct writing of loop equations is Kirchhoff's Voltage Law, which may be restated for our purposes as

Kirchhoff's Voltage Law: the sum of unknown voltage drops in a loop is equal to the sum of known voltage rises in the same loop.

Remember that voltage drop means the current goes from the + terminal to the - terminal. A voltage drop is taken as a negative voltage rise, and viceversa. Using this principle, the equation for a loop is written following the next statement:

Loop equation of loop x: Taking the direction of loop current J_x as reference, the sum of unknown voltage drops is equal to the sum of known voltage rises.

Fig. 7.4(a) shows an isolated loop x to illustrate the principle. V_a and V_b are assumed as unknown voltages. Also, in the direction of J_x, the voltage drop at resistance R_1, which is shared with loop currents J_1 and J_2, the voltage drop is $R_1(J_x - J_1 + J_2)$, as illustrated in inset (b) of the figure. We work similarly with other resistances. Including the black boxes with unknown voltages, and the independent sources, the loop equation for this isolated loop becomes

$$R_1(J_x - J_1 + J_2) + R_2 J_x + R_3(J_x - J_3) + \ldots - V_b + V_a = E_1 - E_2 \quad (7.2)$$

The unknown voltage drops may or may not be related to loop currents. The voltage in current sources are not related, for example, while current dependent voltage sources are related to loop currents.

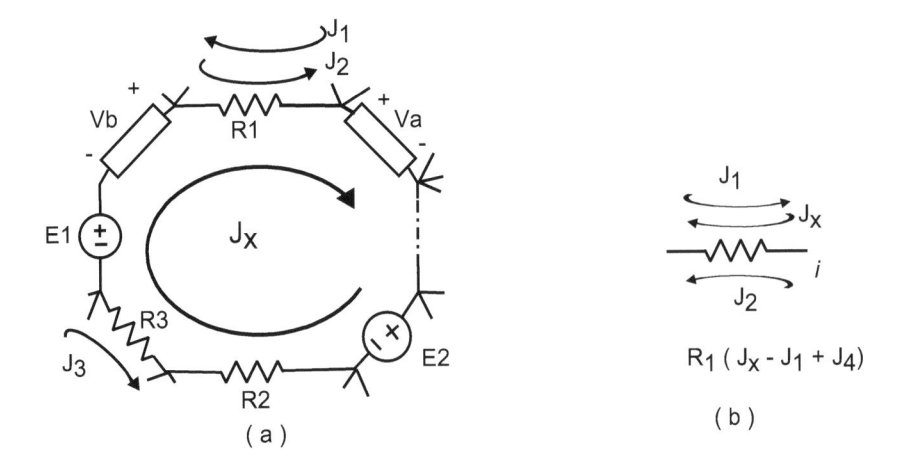

Figure 7.4: (a)Isolated loop x in circuit. (b) Voltage drop at R1 with respect to J_x

7.2 Resistances and voltage sources only

First let us deal with circuits containing only resistances and independent or current dependent voltage sources. Current sources may be transformed either by direct transformation or by first applying shifting of current sources. Our objective for these circuits is to write each loop k equation directly in the form

$$Z_{k1} J_1 + Z_{k2} J_2 + \cdots + Z_{km} J_m + \cdots = V_{Lk} \qquad (7.3)$$

Here

- $J_1, J_2, \ldots$ are the loop currents for loops 1, 2, ...
- Coefficients Z_{kj} are called *loop impedances* or *loop resistances*
- V_{Lk} is the loop voltage of loop k. It is the sum of voltage rises in the loop k

Each loop impedance Z_{mj} will consist of two parts: one due to resistances and another one arising from current-dependent voltage sources:

Z_{mj} = component from R's + component from current dependent voltages

Each component is independent of the other. We deal with these cases separately. The set of all loop equations can be written in matrix form as

$$\mathbf{Zm\,Jm = V_L} \qquad (7.4)$$

7.2. RESISTANCES AND VOLTAGE SOURCES ONLY

7.2.1 Circuits with only resistances and independent voltage sources

From (7.3), and the description above, we see that in the vector of loop voltages $\mathbf{V_L}$, we have

$$V_{Lk} = \sum (\text{known voltage rises in loop } k) \qquad (7.5)$$

Fig. 7.5(b) illustrates how a source E contributes to the vector $\mathbf{VL}$. Let us now consider the resistances.

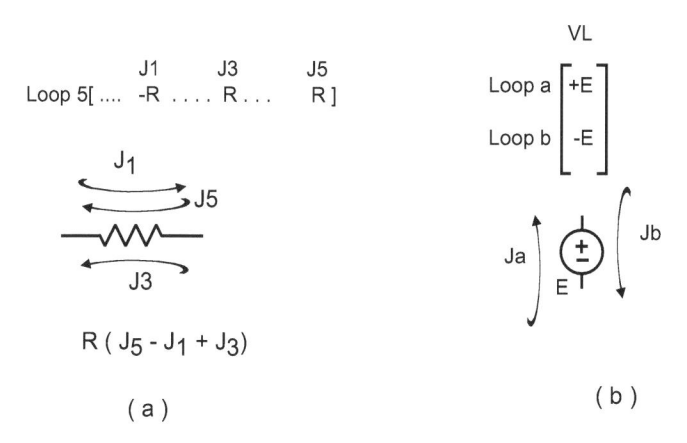

Figure 7.5: (a)Contribution of Resistance to $\mathbf{Zm}$. (b) Contribution of voltage source to $\mathbf{VL}$

Looking at Fig. 7.4(b), we can state the following fact for resistances:

For each resistance included in loop k, and shared with loops $q, p, \ldots$ there will be a voltage drop in the loop of the the form

$$(J_k \pm J_q \pm J_p \pm \ldots)R \qquad (7.6a)$$

Therefore,

$$\text{coefficient of } J_k = \sum \text{resistances in loop } k. \qquad (7.6b)$$

and

$$\text{coefficient of } J_p \text{ is } = \pm \sum \text{resistances shared by loops } k \text{ and } p \qquad (7.6c)$$

where the sign + is taken if both loop currents flow through the resistance in the same direction, and the sign - in the opposite direction.

144   CHAPTER 7. LOOP ANALYSIS

Summarizing, the resistances contribution to coefficients is illustrated by Fig. 7.5(a) on 143, which shows a resistance R contribution to the equation of a loop, labeled 5 for illustration. Mathematically, we write

$$Z_{kj} = \begin{cases} +\sum \text{Resistances in loop } m & \text{for } j = k \\ \pm\sum \text{Resistances shared by loops } m \text{ and } j & \text{for } j \neq m \end{cases} \quad (7.7)$$

Let us bring two examples applying (7.7). In both examples, equations are written in matrix form. The first one goes step by step, showing clearly the application of the criteria. To stress the process, a non numerical example is used. The second one is done straightforward by applying the rules, leaving the step by step verification to the reader.

Example 7.1 *Let us write the loop equations for the circuit of Fig. 7.6, with the three loops selected as shown.*

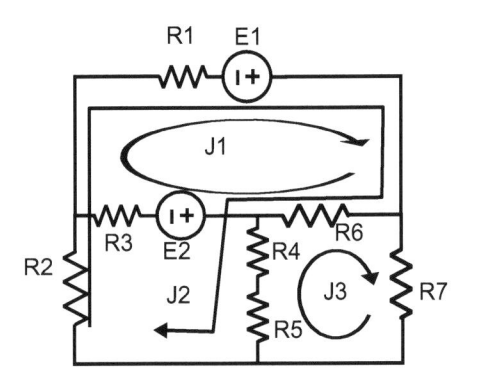

Figure 7.6: Circuit for Step by Step example

To proceed to the individual equations, let us consider the three loops separately, as illustrated in Fig. 7.7.
(1) Equation for Loop 1, shown in detail by inset (a):

(1.1) The resistances contained in loop 1 are R_1, R_3, and R_6. The coefficient for J_1 is therefore $R_1 + R_3 + R_6$.

(1.2) The resistances shared by loop 1 and loop 2 are R_1 and R_6. The respective loop currents go in the same direction through these resistances. Hence, the coefficient for J_2 in this equation is $R_1 + R_6$.

(1.3) The resistance shared by loop 1 and loop 3 is R_6. The respective loop currents go in the opposite direction through this resistance. Hence, the cocfficient for J_3 in this equation is $-R_6$.

(1.VL) Loop 1 contains two independent voltage sources. The direction of J_1 is such that it traverses E_1 in a voltage rise direction, i. e., from the minus

7.2. RESISTANCES AND VOLTAGE SOURCES ONLY

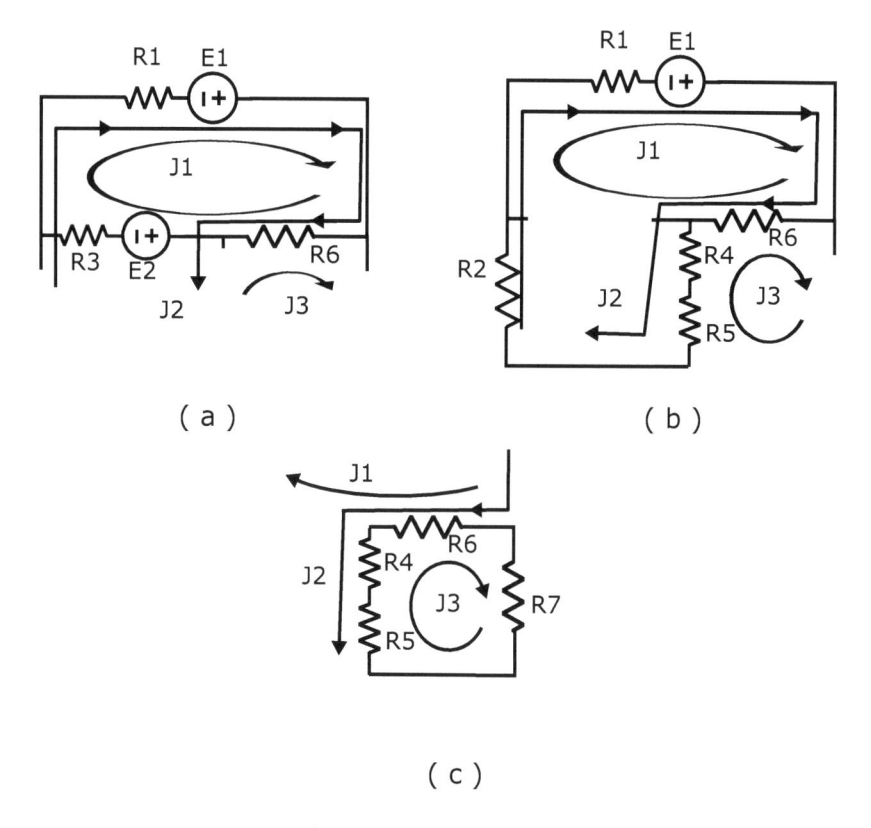

Figure 7.7: (a) Loop 1; (b) Loop 2; (c) Loop 3

(−) to the plus (+) sign. For E_2, the direction is in drop sense. Therefore, the known right hand side should be $E_1 - E_2$.

The equation for loop 1 is therefore

$$(R_1 + R_3 + R_6)\, J_1 + (R_1 + R_6)\, J_2 - R_6\, J_3 = E_1 - E_2 \tag{7.8a}$$

(2) Equation for Loop 2, shown in detail by inset (b):

(2.1) The resistances shared by loop 1 and loop 2 are R_1 and R_6. The respective loop currents go in the same direction through these resistances. Hence, the coefficient for J_1 in this equation is $R_1 + R_6$.

(2.2) The resistances contained in loop 2 are R_1, R_2, R_4, R_5, and R_6. The coefficient for J_2 is therefore $R_1 + R_2 + R_4 + R_5 + R_6$.

(2.3) The resistances shared by loop 2 and loop 3 are R_4, R_5, and R_6. The respective loop currents go in the opposite direction through these resistances.

Hence, the coefficient for J_3 in this equation is $-(R_4 + R_5 + R_6)$.

(2.VL) Loop 2 contains one independent voltage source E_1. The direction of J_2 through E_1 is in a voltage rise sense. Thus, the known right hand side should be E_1.

The equation for loop 2 is then

$$(R_1 + R_6)\, J_1 + (R_1 + R_2 + R_4 + R_5 + R_6)\, J_2 - (R_4 + R_5 + R_6)\, J_3 = E_1 \tag{7.8b}$$

(3) Equation for Loop 3, shown in detail in inset (c):

(3.1) The resistance shared by loop 3 and loop 1 is R_6. The respective loop currents go in the opposite direction through this resistance. Hence, the coefficient for J_1 in this equation is $-R_6$.

(3.2) The resistances shared by loop 3 and loop 2 are R_4, R_5, and R_6. The respective loop currents go in the opposite direction through these resistances. Hence, the coefficient for J_2 in this equation is $-(R_4 + R_5 + R_6)$.

(3.3) The resistances contained in loop 3 are R_7, R_4, R_5, and R_6. The coefficient for J_3 is therefore $R_4 + R_5 + R_6 + R_7$.

(1.VL) Loop 3 contains no independent voltage source. Therefore, the known right hand side is 0.

The equation for loop 2 is therefore

$$-R_6\, J_1 - (R_4 + R_5 + R_6)\, J_2 + (R_4 + R_5 + R_6 + R_7)\, J_3 = 0 \tag{7.8c}$$

In matrix form, we have

$$\begin{array}{c} \\ Loop1 \\ Loop2 \\ Loop3 \end{array} \begin{bmatrix} J_1 & J_2 & J_3 \\ R_1 + R_3 + R_6 & R_1 + R_6 & -R_6 \\ R_1 + R_6 & R_1 + R_2 + R_4 + R_5 + R_6 & -(R_4 + R_5 + R_6) \\ -R_6 & -(R_4 + R_5 + R_6) & R_4 + R_5 + R_6 + R_7 \end{bmatrix}$$

$$\times \begin{bmatrix} J_1 \\ J_2 \\ J_3 \end{bmatrix} = \begin{bmatrix} VL \\ E_1 - E_2 \\ E_1 \\ 0 \end{bmatrix} \tag{7.9}$$

As expected, the coefficient matrix in the example is symmetrical.

7.2. RESISTANCES AND VOLTAGE SOURCES ONLY

Remark: A common mistake among students is the belief that the element voltages and currents depend on the selection and direction of loop currents. The misunderstanding arises from the fact that loop currents are considered as "physical" circuit variables. Although sometimes they have physical meaning, rigourously speaking, they are mathematical entities to find the magnitudes of interest, which are precisely the voltages and currents in the elements. Let us illustrate with the same circuit as before, but now assigning numerical values to elements.

Example 7.2 *Find all the voltage and currents in the circuit of Fig. 7.8 using two different sets of loop currents.*

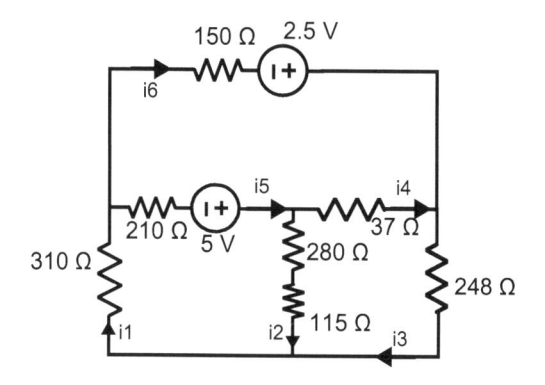

Figure 7.8: Now with numerical values

Fig. 7.9(A) shows as selection of independent loop currents the same set used in the previous example. For the other set, we use mesh currents. This is a planar circuit that has four meshes so any three mesh currents constitute an independent set. We pick the three internal meshes, as shown in Fig. 7.9(B).

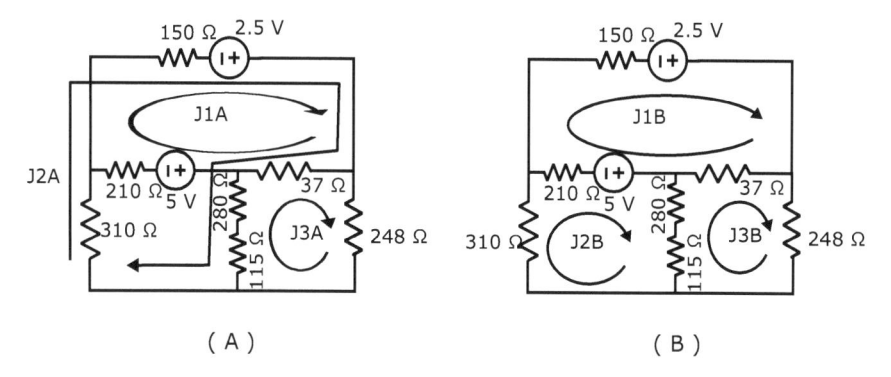

(A)   (B)

Figure 7.9: Two different selections, (A) y (B), of loops.

Before proceeding to establish and solve the loop equations, let us write down how the individual branch currents will be determined in each case. This is done in the following table:

	i_1	i_2	i_3	i_4	i_5	i_6
(A):	J2A	J2A - J3A	J3A	J3A-J1A-J2A	-J1A	J1A + J2A
(B):	J2B	J2B - J3B	J3B	J3B - J1B	J2B - J1B	J1B

In both selections, all currents have physical meaning because in all cases there is an element that is contained in only one loop. From this remark and looking at the table, we expect J2A = J2B and J3A = J3B.

Now let us solve for selection A. The equations are, following the procedure established before,

$$\begin{array}{c} \\ Lp1A \\ Lp2A \\ Lp3A \end{array} \begin{bmatrix} J_{1A} & J_{2A} & J_{3A} \\ 150+37+210 & 150+37 & -37 \\ 150+37 & 310+150+37+280+115 & -(37+280+115) \\ -37 & -(37+280+115) & 280+115+37+248 \end{bmatrix}$$

$$\times \begin{bmatrix} J_{1A} \\ J_{2A} \\ J_{3A} \end{bmatrix} = \begin{bmatrix} V_L \\ 2.5 - 5. \\ 2.5 \\ 0 \end{bmatrix}$$

(7.10)

For the selection B, we have

$$\begin{array}{c} \\ Lp1B \\ Lp2B \\ Lp3B \end{array} \begin{bmatrix} J_{1B} & J_{2B} & J_{3B} \\ 150+37+210 & -210 & -37 \\ -210 & 310+210+280+115 & -(280+115) \\ -37 & -(280+115) & 280+115+37+248 \end{bmatrix}$$

$$\times \begin{bmatrix} J_{1B} \\ J_{2B} \\ J_{3B} \end{bmatrix} = \begin{bmatrix} V_L \\ 2.5 - 5.0 \\ 5.0 \\ 0 \end{bmatrix}$$

(7.11)

7.2. RESISTANCES AND VOLTAGE SOURCES ONLY

Let us store matrices and vectors as za *and* va *for the first set, and* zb *and* vb *for the second. Store solutions as* **xa** *and* **xb**, *respectively Then*

$$\texttt{simult(za,va)} \boxed{\texttt{sto}\rightarrow} \texttt{ xa} \rightarrow \begin{bmatrix} -8.990 \text{ E-3} \\ 6.428 \text{ E-3} \\ 3.595 \text{ E-3} \end{bmatrix}$$

Similarly,

$$\texttt{zb}\wedge \boxed{\texttt{(-)}} 1 * \texttt{vb} \boxed{\texttt{sto}\rightarrow} \texttt{ xb} \rightarrow \begin{bmatrix} -2.562 \text{ E-3} \\ 6.428 \text{ E -3} \\ 3.595 \text{ E-3} \end{bmatrix}$$

As expected, J2A = J2B = 6.428 mA *and* J3A = J3B = 3.59 mA. *Hence, currents* i_1, i_2, *and* i_3 *have the same value in both systems. Checking for the other cases:*

For i_4: `xa[3,1] - xa[1,1] - xa[2,1]` ENTER → 6.157 E-3
and `xb[3,1] - xb[1,1]` ENTER → 6.157 E-3, i. e., i_4 = 6.157 mA.

For i_5: `xb[2,1] - xb[1,1]` ENTER → 8.990 E-3, i. e., i_5 = 8.990 mA., *same as* -`xa[1,1]` (-J1A)

For i_6: `xa[1,1] + xa[2,1]` ENTER → -2.562 E-3, i. e., i_6 = -2.562 mA., *same as* `xb[1,1]` (J1B)

Scaling units

We can introduce the resistances in kilohms units, and then all currents can be interpreted in miliampere units.

7.2.2 Current-controlled voltage sources

Let us now introduce current controlled voltage sources into the game. For a voltage controlled one, with the controlling voltage related to current, then we apply this relationship for conversion.

The general situation is depicted in Fig. 7.10. Here, without loss of generality, the source is shown belonging to two loops only, p and q. Similarly, the controlling current is expressed here as the difference of two loop currents as Ja - Jb.

The equations that will be modified by the source are for those loops to which the the source belongs. The way in which this modification takes place is illustrated in the same figure. Namely, the transresistance r will add to or subtract from the coefficients of the respective controlling loop currents.

150   CHAPTER 7. LOOP ANALYSIS

```
              Jp                        Ja      Jb
                              Loop p [ ...  r  .....  -r ...]
              Jq
                                        Ja      Jb
                              Loop q [ ...  -r  .....  r ...]
              r(Ja - Jb)
```

Figure 7.10: CCVS affects the loop equations in which it is embedded

We illustrate this remark using the situation shown in Fig. 7.10. For loop p, seeing that current Jp traverses the source in the voltage drop direction, and thus positive, we write the equation in the form

$$R_{p1} J_1 + \ldots + R_{pa} J_a + \ldots + R_{pb} J_b + \ldots + r(J_a - J_b) = VL_p \quad (7.12a)$$

where the coefficient R_{pk} emphasizes that only resistances have been considered to build them, with the rules of the previous section. Similarly, for loop q,

$$R_{q1} J_1 + \ldots + R_{qa} J_a + \ldots + R_{qb} J_b + \ldots - r(J_a - J_b) = VL_q \quad (7.12b)$$

Rearranging the equations with Algebra,

$$R_{p1} J_1 + \ldots + (R_{pa} + r) J_a + \ldots + (R_{pb} - r) J_b + \ldots = VL_p \quad (7.13a)$$

$$R_{q1} J_1 + \ldots + (R_{qa} - r) J_a + \ldots + (R_{qb} + r) J_b + \ldots = VL_q \quad (7.13b)$$

When you set up the equations, you may do it in two steps, starting with the set (7.12), or directly to the final forms (7.13).

Example 7.3 *Let us find the voltage Vo in the circuit of Fig. 7.11 using mesh analysis.*

We proceed in two steps. First, imagine that the dependent source is not included, considering it to be a short circuit, and set up the matrices. With the dependent source subtituted with a short circuit, **Zm** *– with "R_{ij}"– is*

	J_1	J_2	J_3	J_4
Lp1	$10000 + 600$	-10000	-600	0
Lp2	-10000	$10000 + 500 + 1200$	-1200	0
Lp3	-600	-1200	$600 + 1200 + 150$	-150
Lp4	0	0	-150	$150 + 1500$

and $\mathbf{VL} = [3 \ 0 \ 0 \ 0]^\mathrm{T}$.

7.2. RESISTANCES AND VOLTAGE SOURCES ONLY

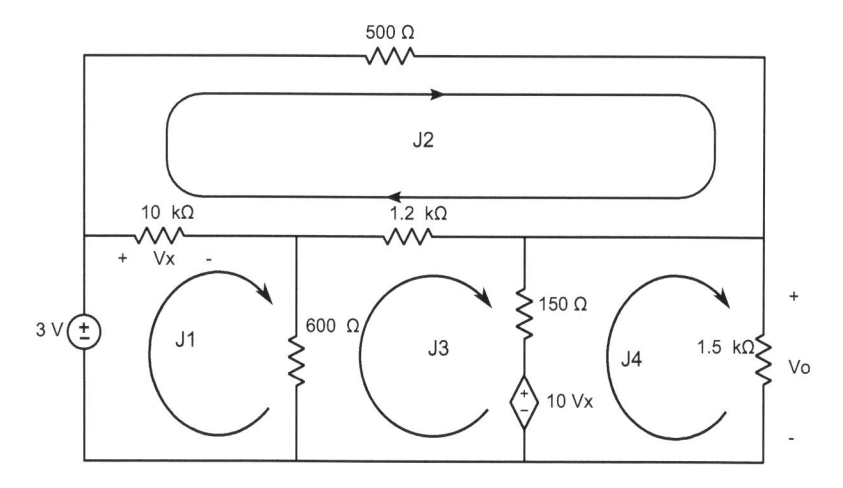

Figure 7.11: Example for Mesh Analysis with Current Controlle Voltage Source.

Let us store the matrices as **zm** and **vl** in our calculator. We are now ready to consider the dependent voltage source. This one is actually a voltage controlled source. But since Vx = 10000 (J1-J2), we can express it in terms of loop currents as 10(10E3)(J1-J2)=10E4(J1 - J2).

Looking at the circuit, we see that this dependent source is in drop sense for J3 and rise direction for J4. Using as reference what has been said before, in equations of J3 and J4, the coefficients for J1 and J2 will be affected as shown in (7.13).

Therefore, the final matrix is as shown below, where I put the source contribution in bold, and using E notation:

Zm =

	J_1	J_2	J_3	J_4
Lp1	10000 + 600	−10000	−600	0
Lp2	−10000	10000 + 500 + 1200	−1200	0
Lp3	**−600 + 10E4**	**−1200 − 10E4**	600 + 1200 + 150	−150
Lp4	**−10E4**	**10E4**	−150	150 + 1500

In the calculator, having already the matrix without dependent source, we proceed to get the final matrix **zm** by editing the following four entries on the command line:

zm[3,1] + 10E4 [sto→] zm[3,1] → 99400.

zm[3,2] - 10E4 [sto→] zm[3,2] → -1.012E5

zm[4,1] - 10E4 [sto→] zm[4,1] → -1.E5

zm[4,2] + 10E4 [sto→] zm[4,2] → 1.E5

We are now ready to find the solution as

$$\text{simult(zm,vl)} \; \boxed{\text{ENTER}} \; \rightarrow \begin{bmatrix} -6.61\text{E-}3 \\ -6.69\text{E-}3 \\ -10.19\text{E-}3 \\ 4.23\text{E-}3 \end{bmatrix}$$

We have therefore $J4 = 4.23$ mA. *We multiply now this current by the resistance* 1.5 kΩ *to find the desired output voltage:*

ans[4,1]*1.5E3 [ENTER] → 6.3455E0

Hence, Vo = 6.35 V.

A trick: Insert equation or change variable You are working with technology, a calculator. Why not include more information to do things faster? In the previous example, since $J4 = V_o/1500$, we could have change variable directly on the equations. Thus, our matrix, *including the CCVS*, would be

Zm =

	J_1	J_2	J_3	V_o
L1	$10000 + 600$	-10000	-600	0
L2	-10000	$10000 + 500 + 1200$	-1200	0
L3	$-600 + 100E3$	$-1200 - 100E3$	$600 + 1200 + 150$	$\frac{-150}{1500}$
L4	$0 - 100E3$	$0 + 100E3$	-150	$\frac{150+1500}{1500}$

and the result would be direct. Try it!

7.2.3 Indefinite mesh matrix

Let us limit ourselves to planar networks, and select all meshes as loops, including the external mesh. Furthermore, let all internal meshes be clockwise oriented, and the external mesh counterclockwise oriented. The resultant matrix **Zm** is called *indefinite mesh impedance matrix* (IMM). By extension,

7.3. PROGRAMMING MESH EQUATIONS I

Vm, may be called indefinite mesh voltage vector. The IMM has many uses[1]. Here, the purpose is to use it to program mesh analysis.

Both **Zm** and **Vm** have the characteristic that the sum of the rows are 0. Moreover, in **Zm**, the sum of columns is also 0. As a consequence, its determinant is 0 an the set of equations does not have a solution. For our purpose, however, an important characteristic is that we can delete the column and row that correspond to a mesh in **Zm** and **Vm**. The system that results corresponds to an independent set of meshes, those that remain.

7.3 Programming mesh equations I

Let us program to generate the matrices **Zm** and **Vm** and solve for the mesh current vector **Jm** . The basic algorithm for circuits with characteristics mentioned so far, that is, with resistances, voltage sources and current controlled voltage sources, is described in the following steps using indefinite mesh matrix. The mesh we want to drop is denoted as "N+1" for easy row and column deletion.

7.3.1 Pseudocode for program

The pseudo code given below assumes that you have at least one loop in your system.

1. **Number of meshes** Specify number of meshes N, excluding the one you drop. Denote the dropped mesh as N+1.
2. **Initialization** Initialize matrix **Zm** = 0 of order (N+1)x(N+1) as well as vector **Vm** of order (N+1)x1
3. **Resistance Subroutine** For each resistance of value R ($\neq 0, \infty$) belonging to meshes j and k do:

    1. $Zm[j,j] = Zm[j,j] + R$
    2. $Zm[k,k] = Zm[j,j] + R$
    3. $Zm[j,k] = Zm[j,k] - R$
    4. $Zm[k,j] = Zm[k,j] - R$

4. **Current controlled voltage source subroutine** If there are CCVS's, then, for each source with transresistance rm in drop direction for mesh j and rise direction for mesh k, and controlled by $(J_p - J_q)$ do:

    1. $Zm[j,p] = Zm[j,p] + rm$
    2. $Zm[j,q] = Zm[j,q] - rm$
    3. $Zm[k,p] = Zm[k,p] - rm$

---

[1] For further information the reader may consult the paper by Kiss, W.F. ; Gilson, R.A. , "*On the formulation of the indefinite matrix*", IEEE J. of Solid-State Circuits, vol. 3, No. 3, pp. 307-308, 1968

4. $Zm[k,q] = Zm[k,q] + rm$

5. Creating definite Zm: Delete both row and column (N+1) of **Zm**.

6. Voltage source Subroutine For each cource source of value V ($\neq 0, \infty$) in drop position for mesh j and rise position for mesh k do:

1. $Vm[j,1] = Vm[j,1] - V$
2. $Vm[k,1] = Vm[k,1] + V$

7. Creating definite Vm: Delete row (N+1) of **Vm**

8. Solve for Jm: If $|\mathbf{Zm}| \neq 0$ then $\mathbf{Jm} = \mathbf{Zm}^{-1}\mathbf{Vm}$

If there are no voltage sources, you can use the program to get the matrix **Zm**. Let us now use the above steps to develop programs.

7.3.2 Preparing input for the program

The program receives the following inputs. Remember, you must include the outer mesh current in counter clockwise direction, and all internal mesh currents in clockwise direction.

1. **n**: A non zero number of independent meshes. The description includes all meshes; call the one to be dropped mesh "n+1" in the following matrices

2. An $m \times 3$ matrix **R** for the m resistances in the circuit. Each row k consists of three items:
   - Element $(k, 1)$ is a non-zero and non-infinite resistance value,
   - Elements $(k, 2)$ and $(k, 3)$ are the two meshes to which it belongs.

3. Variable **vs**. This variable is 0 if there are no independent voltage sources, which is the case if you are only interested on **Zm**. Otherwise, is a $p \times 3$ matrix **vs** for the p voltage sources in the circuit. Each row **k** consists of three items:
   - Element $(k, 1)$ is the finite voltage source value. It may be 0.
   - Element $(k, 2)$ is the mesh number where the current is in voltage drop sense through the source,
   - Element $(k, 3)$ is the mesh number where the current is in voltage rise sense through the source.

4. Input **rs**. This value is 0 if the circuit contains no current controlled voltage sources. Otherwise, it is a $p \times 5$ matrix. The source $rm(Jp-Jq)$ is defined in row k as follows:
   - Element $(k, 1)$ is a finite transresistance rm value.

7.3. PROGRAMMING MESH EQUATIONS I

- Element $(k, 2)$ is the mesh where current is in voltage drop sense (entering through the + terminal),
- Element $(k, 3)$ is the mesh where current is in voltage rise sense (entering through the - terminal),
- Element $(k, 4)$ is the mesh number for Jp.
- Element $(k, 5)$ is the mesh number for Jq.

Fig. 7.12 illustrates the rows for each of the input matrices,

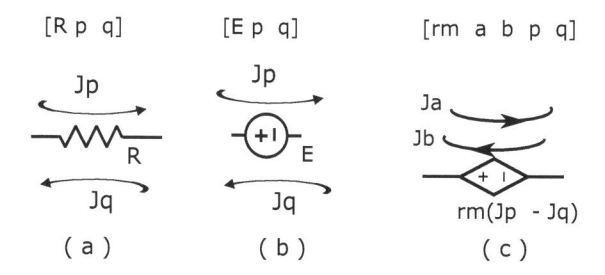

Figure 7.12: Illustrating rows for matrices (a) r, (b) vs, and (c) rs

7.3.3 Program for the calculators

Fig. 7.13 shows a program in these environments. Lines are commented on the right column which are not part of the program, although you can include them if you wish. Except for the case of checking if the circuit description is valid, I have not attempted any extra feature like data checking or other advisable programming precautions for a general case. I wrote the program for personal use and I assume I will not provide wrong data.

The program returns three matrices: The loop impedance matrix zm, the loop voltage vector vm, and the vector of loop currentes jm. These results are in memory once you run the program.

Example 7.4 *Let us illustrate using the circuit worked in example 7.3. In this circuit the four inner meshes were selected. This means that the extra mesh is the outer mesh, which goes in the counterclock direction. This is illustrated in Fig. 7.14 on page 157.*

The inputs for the program to solve for this circuit, with the selection shown, are as follows. Notice that when transforming the voltage source to current controlled, $10Vx = 100,000(J_1 - J_3)$:

```
 1  loop1(n,r,vs,rs)
 2  Pgrm
 3  Local a,b,c,d,e,t
 4  newMat(n+1,n+1)→ zm           Initialize indefinite mesh matrix
 5  newMat(n+1,1)→ vm             and indefinite Vm vector
 6  For t,1,rowDim(r)             Start Zm with resistances
 7  r[t,1]→a                      a=R;
 8  r[t,2]→b:r[t,3]→c             R belongs to meshes b and c
 9  zm[b,b]+a→zm[b,b]             Updating $Zm_{bb}$
10  zm[c,c]+a→zm[c,c]             Update $Zm_{cc}$
11  zm[b,c]-a→zm[b,c]             Updating $Zm_{bc}$ and
12  zm[c,b]-a→zm[c,b]             Updating $Zm_{cb}$
13  EndFor                        End Resistance subroutine
14                                Blank line for easy reading
15  If rs/=0 Then                 Case when there are VCCS's
16  For t,1,rowDim(rs)            Start introduction of dependent sources
17  rs[t,1]→a                     For simpler notation in writing: a=rm;
18  rs[t,2]→b:rs[t,3]→c           drop in mesh b, raise in mesh c
19  rs[t,4]→d:  rs[t,5]→e         controlled by Jmd - Jme
20  zm[b,d]+a→zm[b,d]             Updating $Zm_{bd}$ and
21  zm[b,e]-a→zm[b,e]             Updating $Zm_{be}$ and
22  zm[c,d]-a→zm[c,d]             Updating $Zm_{cd}$ and
23  zm[c,e]+a→zm[c,e]             Updating $Zm_{ce}$ and
24  EndFor                        End reading rs
25  EndIf                         End case for dependent sources
26  subMat(zm,1,1,n,n)→ zm        Create definite zm
27                                Blank line for easy reading
28  If vs/=0 Then                 Case for independent sources
29  For t,1,rowDim(vs)            Introduce independent sources
30  vs[t,1]→a                     For simpler notation in writing: a=I;
31  vs[t,2]→b:vs[t,3]→c           drop in mesh b, raise in mesh c
32  vm[b,1]-a→vm[b,1]             Updating $Vm_b$
33  vm[c,1]+a→vm[c,1]             Updating $Vm_c$
34  EndFor                        End adding independent sources
35  subMat(vm,1,1,n,1)→ vm        Definite vm
36  If det(zm)/=0 Then            If circuit is valid
37  zm∧-1*im→ jm                  Solving for mesh current vector jm
38  Else                          Otherwise
39  Disp ''Description non valid''  Message for invalid description
40  EndIf                         End mesh currents
41  EndIf                         End case of independent sources
42  EndPrgm                       Exit Program
```

Figure 7.13: Example of a program for mesh analysis (resistances and current sources). Remember that → in the program stands for sto→

$$n = 4; \quad \mathbf{r} = \begin{bmatrix} 10E3 & 1 & 2 \\ 600 & 1 & 3 \\ 1200 & 2 & 3 \\ 150 & 3 & 4 \\ 1500 & 4 & 5 \\ 500 & 2 & 5 \end{bmatrix}; \quad \mathbf{vs} = [3,5,1]; \quad \mathbf{rs} = [100E3,3,4,1,2]$$

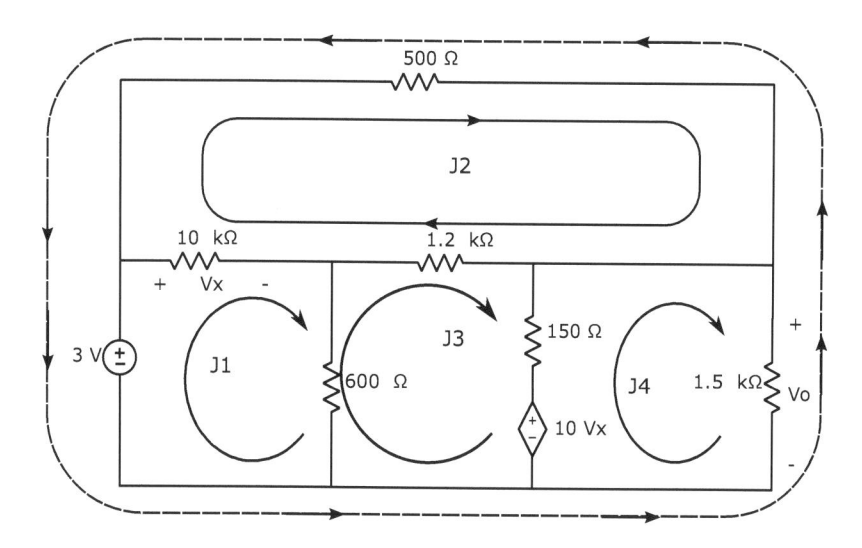

Figure 7.14: Circuit to illustrate inputs to program `mesh1`.

7.4 Loop analysis with current sources

When there are current sources in your circuit you will have unknown voltages which do not depend on your loop currents. Therefore, the number of variables increases. Each current source introduces an unknown voltage and also an equation or a known value and the system is solvable[2].

You have two options. Either you take a reduced number of equations not involving the unknown voltages, or else you include those voltages in the system. The first option is the preferred one when solving with pencil and paper. Since we are using calculators which help us solve the equations easily, the way we proceed is a matter of choice.

7.4.1 Modified Loop Analysis

When you want all the unknown voltages as well, you can include them in the equations. You don't need to select loops in a special way. Just do it the way you prefer! The result has been called by some authors modified loop analysis.

The following example by hand will illustrate the general method. Remember: *In the unknown side, an unknown voltage drop has coefficient +1 (positive) and an unknown voltage rise has coefficient -1 (negative)!*

Example 7.5 *To illustrate the process, the circuit in Fig. 7.15 with meshes as usually taken by students. In this circuit, let us find the power generated by each of the three sources.*

---
[2]Remember that we always assume well stated systems for our purposes.

158   CHAPTER 7. LOOP ANALYSIS

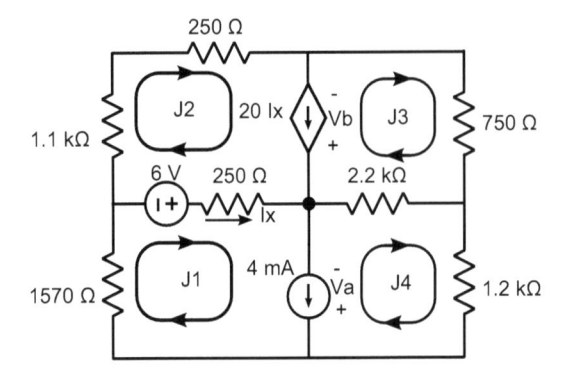

Figure 7.15: Modified mesh analysis for the circuit of Fig. 7.16

Before proceeding to set up equations, we need to relate the loop currents and the unknown voltages with the requested powers. Moreover, we need the equations introduced by the current sources.

The reader can verify that the power calculations can be written in terms of the variables in the equations as

$$P_{V-source} = 6 \times (J_1 - J_2); \quad P_{I-source} = (4 \times 10^{-3})V_a; \quad P_{CCCS} = V_b \times (J_2 - J_3) \tag{7.14}$$

As for the other issue, the equations defined by the current sources are:

$$J_1 - J_4 = 4 \times 10^{-3}; \text{ and } J_2 - J_3 = 20(J_1 - J_2)$$

The latter can be rearranged as $-20 J_1 + 21 J_2 - J_3 = 0$ *Now, set up the equations. The matrices involved are as follows:*

Zm =

	J_1	J_2	J_3	J_4	V_a	V_b
Lp1	$250+1570$	-250	0	0	-1	0
Lp2	-250	$1100+250+250$	0	0	0	-1
Lp3	0	0	$2200+750$	-2200	0	1
Lp4	0	0	-2200	$2200+1200$	1	0
CCCS	-20	21	-1	0	0	0
Isrc	1	0	0	-1	0	0

(7.15)

$$\mathbf{VL} = \begin{bmatrix} 6 & -6 & 0 & 0 & 0 & 0.004 \end{bmatrix}^T \tag{7.16}$$

7.4. LOOP ANALYSIS WITH CURRENT SOURCES

Now solve the system, either with `simult(zm,vl)` *or using the multiplication* `zm\`(-)`1* vl`. *Store the result in* `xm`. *For easiness, we show the result in transposed form:*

`xm= [372.77E-6   265.86E-6  -1.873E-3  372.77E-6 -5.388E-0 6.3454E-0]`^{T}

The vector elements are are, respectively, J_1, J_2, J_3, J_4, V_a, *and* V_b. *We can now solve for the required power calculations of* (7.14).

For $P_{V-source} = 6 \times (J_1 - J_2)$: `6 * (xm[1,1] - xm[1,2])` → `641.65E-6`
For $P_{I-source} = (4 \times 10^{-3})V_a$: `4E-3 * xm[1,5]` → `-21.552E-3`
For $P_{CCCS} = V_b \times (J_2 - J_3)$: `xm[1,6] * (xm[1,2] - xm[1,3])` → `13.572E-3`

Therefore, the voltage source generates 641.65 µW *and the dependent source* 13.572 mW, *but the independent current source absorbs* 21.552 mW.

Remarks on the system of equations

Observing the previous example, it can be seen that the equations have the form **Zmla Jma = Vmla**, where matrices and vectors can be partitioned, so that the final form is

$$\begin{bmatrix} \mathbf{Z_m} & \mathbf{Vx} \\ \mathbf{IEQ} & 0 \end{bmatrix} \begin{bmatrix} \mathbf{J}_m \\ \mathbf{Vxs} \end{bmatrix} = \begin{bmatrix} VL \\ \text{Isval} \end{bmatrix} \qquad (7.17)$$

where, considering N the number of independent meshes and B the number of current sources,

Zm of order N x N, **VL** of order N x 1 are the same matrices already introduced.

Jm is the vector of unknown mesh currents

Vx is an N x B matrix of for the unknown voltages in sources, and **IEQ** is the B x N matrix defining the current sources equations. The unknown voltage in column j of **Vx** stands for that of the current for row j of **IEQ**. This colum j has +1 in the row for mesh current J_p if the current runs in voltage drop direction, -1 if it runs in voltage rise direction, and 0 otherwise.

The system **IEQ Vxs = Isval** is the set of equations for current sources.

The system is programmable, following similar steps to those for modified nodal analysis. I leave to the interested reader the exercise of writing the program.

7.4.2 Reducing the number of equations

This method is based on the principle, already stated before, that if an unknown variable appears in only one equation, than the equation may be taken separately. If the variable is not of interest, the equation is not solved.

In the case of loop equations, the isolation of the unknown voltage is achieved by selecting loops in such a way that one and only one loop goes through the current source. If the source is independent, we can take the loop current as a known value[3]. If it is dependent, then we need to add the equation of the source to the system.

Let us look at an example.

Example 7.6 *Let us set up the equations for the circuit from example 7.5, reproduced in Fig. 7.16 without the meshes. This circuit contains two current sources. Our objective is to write a set of equations which does not include the voltages at those sources, V_a and V_b.*

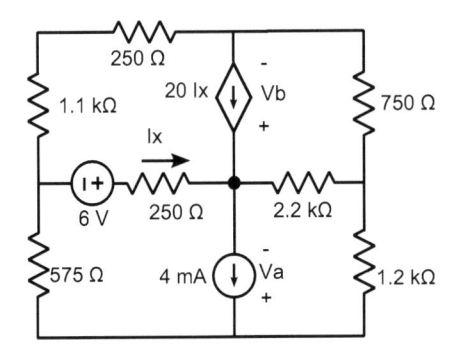

Figure 7.16: Circuit with current sources

We start by drawing loops in a subcircuit where the current sources have been removed, as shown in Fig. 7.17(a) on next page. I have chosen J2 as the outer mesh because in that way I have $Ix = J1$. However, this has been a choice, it does not have to be that way.

Next, sources are reinserted one at a time. Each current source will define at least one loop containing the source and with all other elements not a current source. One possible selection is the one shown in Fig. 7.17(b), where in addition I have taken care not to have Ix depending on two current loops.

With this selection, we see that $J4 = 4$ mA is a known value, and $J3 = 20\,Ix = 20J1$ defines an equation. The equations, in expanded form, are then

Loop 1:

$$(250+2200+1200+575)\,J_1 - (1200+575)\,J_2 + 2200\,J3 = 6 + (2200+1200)(4\times 10^{-3}) \tag{7.18a}$$

Loop 2

$$-(1200+575)\,J_1 + (1200+575+250+750)\,J_2 + 750\,J3 = -1200(4\times 10^{-3}) \tag{7.18b}$$

---

[3]This applies equally to symbolic cases, applying the principle of proportionality

7.4. LOOP ANALYSIS WITH CURRENT SOURCES

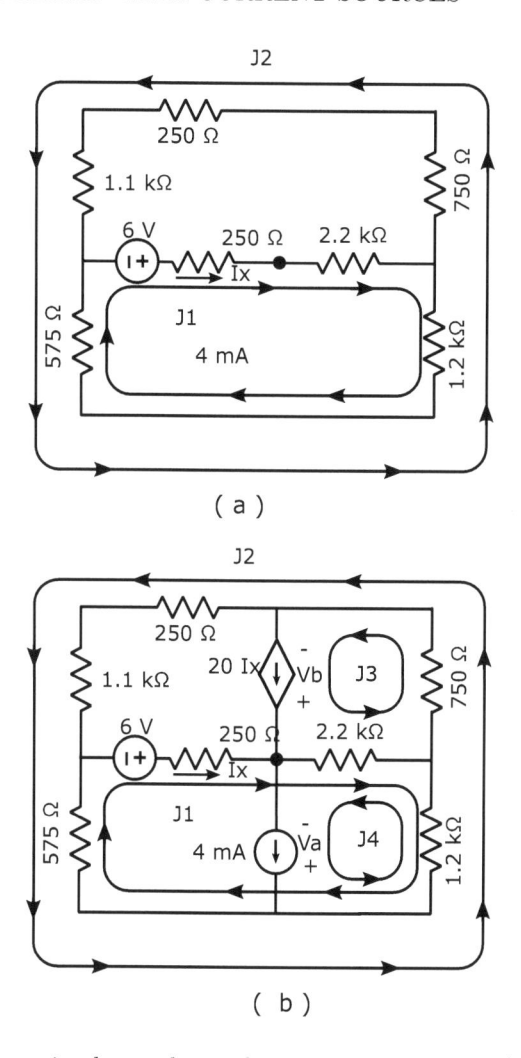

Figure 7.17: Steps in the analysis of circuit in Fig. 7.16: a) step 1: ignore current sources; b) step2 include them, one loop per source

Dependent current Source

$$-20 J_1 + J_3 = 0 \qquad (7.18c)$$

The current sources' voltages are not included in these three equations. If any voltage is needed, add the equation of the corresponding loop.

For planar circuits, this method has been called super-mesh method. Again, in my opinion this notation has been included ignoring previous methods of loop selection. One, for example is using trees, and selecting the tree in such a way that no source current is included in it. The loops that result are called fundamental loops.

7.4.3 Hybrid systems

In many situations, it may become necessary to consider the voltage in current sources. That means that we do not eliminate the loops that go through that source. To consider an example, take the problem from example 7.6. If for any reason we need the voltage of the dependent current source, then include in the set of equations (7.18) that for loop 3:

$$2.2E3 J_1 + 2.2E3 J_2 + (2.2E3 + 750) J_3 + 750 J_4 - V_b = 0$$

You have now four equations with four variables, but you solve for the voltage in just one step.

7.5 Superposition and symbolic sources

Just as it has been done in previous chapters, you deal with symbolic sources using a value 1 for it in the equations, and multiplying by the variable in the final interpretation. Similarly, you separate the known side in columns for superposition, or several sources of the form $Kf(t)$ or several symbolic sources.

To illustrate, in example 7.2, if you want to verify that the contribution of each source is independent of the loop selection, use matrices

$$\mathbf{V_{LA}} = \begin{bmatrix} 2.5 & -5. \\ 2.5 & 0 \\ 0 & 0 \end{bmatrix} \text{ and } \mathbf{V_{LB}} = \begin{bmatrix} 2.5 & -5.0 \\ 0 & 5.0 \\ 0 & 0 \end{bmatrix}$$

7.6 Closing remark

Remember, being able to state and solve the equations is only part of the game. A very important part, because without this skill there is nothing else to do But what makes the road interesting, are the applications you are able to do with the results. In this sense, I think that the use of the calculator enhances its value as a tool, because not only it provides us a tool to solve equations, but with appropriate considerations in the process, we can deduce interpretations for applications directly from the solution.

Let's continue!

CHAPTER 8

Network Functions

Network functions provide us with information about what we should expect as a response for a given input signal, without necessarily knowing the details of what is happening inside the circuit. By input signal, we refer to a voltage, current, or power present at the port called *input port*. The response may be the voltage or current at an element, which may be an open or short circuit, or the power absorbed. The element is sometimes called *load*, and the response is usually called *output*.

This chapter deals with finding the network functions of a circuit. This is achieved through a proper interpretation of results from analysis using any of the methods already discussed. Or any other that the reader may prefer. Sometimes additional equations may be introduced in the set for convenience.

8.1 Definition of the network functions

This chapter is focused on circuits with one independent source, which provides the *input signal*, either voltage or current. The two terminals to which the source is connected define the input port. The signal of interest, the *output signal*, may be located at the input port or else it may be a voltage or current at another element called load, which may also be an open circuit or a short-circuit. This situation is illustrated by Fig. 8.1. Notice that the input current I_{in} enters the circuit through the terminal connected to the plus sign of the voltage V_{in}. The input power $P_{in} = V_{in} I_{in}$ is therefore the power generated by the source and absorbed by the circuit.

In general, denote the input and output signals as X_{in} and X_{out}, without any particular reference to the type of magnitude, except if confusion may arise. The output signal could be the current or voltage at the source. In this case, if the input signal is the voltage of the source, then the input current would be the output signal. And the other way around. When both the

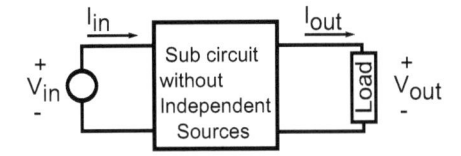

Figure 8.1: One source (input) and load (output) circuit

input and the output signal are at the input port, the network function is called *port function*. Otherwise, it is a *transfer function*.

As far as the process of analysis concerns, the input signal source may be of any type. If it is a voltage source, then v_{in} is known and i_{in} is to be calculated, and vice versa. In fact, except for situations where a mathematical inconsistency may arise, the treatment of these magnitudes is a matter of mathematical convenience.

Because of the principle of proportionality discussed in chapter 4, we know that the output signal will be proportional to the input signal. That is, $X_{out} = K X_{in}$, where K is a constant that depends completely on the circuit elements and the topology. In other words, K is defined entirely by the circuit and is independent of the input signal. This allows us to define a *transfer function*, or more accurately, a *network function* with an expression of the form

$$K = \frac{X_{out}}{X_{in}} \quad (8.1)$$

When the input and output signals are defined, then they receive a particular name. The different functions are the following, all making reference to Fig. 8.1:

Equivalent Resistance $\quad R_{eq} = \dfrac{V_{in}}{I_{in}} \quad$ Equivalent Conductance $\quad G_{eq} = \dfrac{I_{in}}{V_{in}}$

Current transfer $\quad A_I = \dfrac{I_{out}}{I_{in}} \quad$ Voltage transfer $\quad A_v = \dfrac{V_{out}}{V_{in}}$

Transresistance $\quad R_m = \dfrac{V_{out}}{I_{in}} \quad$ Transconductance $\quad G_m = \dfrac{I_{out}}{V_{in}}$

In specific applications, these functions are called with other names. For example, what we have called voltage and current transfer functions, in electronics are voltage and current gains, respectively. When gain is less than 1, the term *attenuation* is also used. Similarly, the equivalent resistance may be called input or output resistance, depending on the use or name given to a pair of terminals in an application. These and other notations are specific to a field or application, but refer always to the same definition.

The power transfer function, or power gain, is defined as

8.2. FINDING THE NETWORK FUNCTIONS

$$A_p = \frac{P_{out}}{P_{in}} = \frac{V_{out} I_{out}}{V_{in} I_{in}} \qquad (8.2)$$

Again, notice this is the ratio of power absorbed to the power delivered by the source.

8.2 Finding the network functions

Even though finding a network function is just a matter of interpreting results, some students have problems relating the analysis methods to the goal of calculating a network function. In an attempt to explain this relationship, let me break the process in two parts. First, an example working with reduction and transformations, so we can use all methods but be sure to plan. Then, let me lay out basic general principles and illustrate with examples using the other methods.

8.2.1 Using reduction and transformation

As you might realize, reduction of a circuit to a single equivalent resistance is an example of how to find this port function or its inverse, the equivalent conductance. Similarly, the voltage and current divider formulas are examples for voltage and current transfer functions, respectively, of particular circuits.

Let us now consider another examples of calculation to illustrate some general remarks. To further point out the difference in strategy, let us work with a circuit used previously.

Some general remarks for strategy could be the following:

a) Reduce the circuit to the simplest possible topology without loosing the variables of interest.

b) From the above step, you might need to further transform or reduce the circuit, hiding variables of interest. In that case, before proceeding, set up the equations or steps you will need to recover the information.

c) If your are calculating several functions, try to gather your different magnitudes in groups such as lists or vectors, so you can compute easily and also make it easier to recall the values.

Example 8.1 *Let us take the circuit of example 5.5 on page 61, with the 20 V source as the input source, and the 605 Ω resistance as the load. As a reference to the example, with respect to the notations in Fig. 8.2(a), we see that I_6 is the input current, I_{in}, and V_4 and I_4 are the output voltage and current, Vo, Io, respectively. Our interest is to find <u>all</u> functions. Hence, let us create the list L6 for all the variables of interest, with the meaning L6 = { $V_{in}, I_{in}, I_{out}, V_{out}$ }. Thus, our first step:*

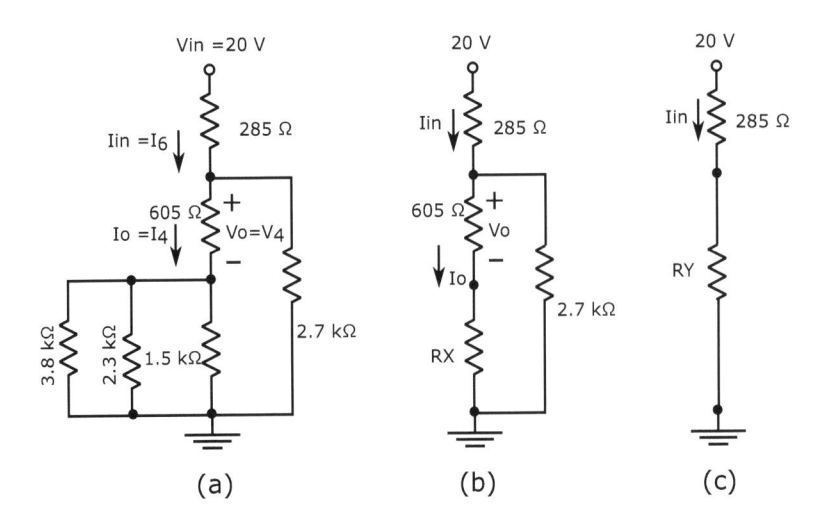

Figure 8.2: Circuit from example 5.5

Initialize L6: $\{20,0,0,0\}$ sto→ L6 enter

Since we are not interested in other magnitudes, we can reduce the circuit to that in (b), which is the maximum reduction we can have with still all the magnitudes of interest visible. Here, Rx = 38 kΩ|| 2.3 kΩ|| 1.5 kΩ. *Hence, our second step:*

pl({1.5 EE 3, 2.3 EE 3, 3.8 EE 3}) sto→ X enter → 732.81E0 (Rx = 732.81 Ω)

From this inset b, we find the relations

$$Io = \frac{Iin \times 2700}{2700 + 605 + Rx} \text{ and } Vo = 605\, Io$$

which we use to find Io and Vo, once Iin is calculated. Even if the circuit at this step is simple enough to find the input current, let us reduce it one more step, to Fig. 8.2(c):

pl({X + 605, 2.7 EE 3}) sto→ Y enter → 894.57E0 (Ry = 894.57 Ω)

Now calculate Iin and store as L6[2]:

20/(285 + Y) sto→ L6[2] enter → 16.955E-3 (I_6 = 16.96 mA)

Find Io and store as L6[3]:

8.2. FINDING THE NETWORK FUNCTIONS

× 2.7 [EE] 3/sum(L2) [sto→] L6[3] [enter] → 11.348E-3E-3

Find Vo and store as L6[4]:

× 605 [sto→] L6[4] [ENTER] → 6.865E0 (Vo= 6.87 V).

Next, L6/L6[1] → {1.E0, 845.45E-6, 567E-6, 343.27E-3}
shows the numerical values of $\{1, G_{eq}, G_m, A_v\}$, respectively.

Now L6/L6[2] → {1.1828E3, 1.E0, 671.12E-3, 406.02E-3}
shows the numerical values of $\{R_{eq}, 1, A_I, R_m\}$, respectively.

For the power transfer, $A_p = (V_{out} I_{out})/(V_{in} I_{in})$, do

L6[3]*L6[4]/(L6[1]*L6[2]) → 230.38E-3

As you can see from this example, the functions have been obtained by *proper interpretation* of the results. The value of 20 V for the source is irrelevant as far as the functions concern. We could have used a 1 V source, and the first set of results could have been directly obtained.

In transformation and reduction methods, as illustrated above, one should be aware of each step and each reduction. What is lost, what is kept. And, of course, practice and practice so you develop your intuition and speed up your strategy planning.

Before proceeding further let's make a brief detour to facilitate numbers and readings:

Scaling: If all resistances are provided in kΩ units (conductances in mS units), then the equivalent resistance R_{eq} and trans-resistance R_m functions are in kΩ units too, while the equivalent conductance G_{eq} and trans-conductance G_m functions are in mS units. The voltage and current transfer functions, A_v, A_I, as well as the power transfer A_P remain unaltered.

8.2.2 General principles of calculation

You are going to find a network function by proceeding with an analysis of the circuit, by whatever method you prefer. But to simplify your task, spend few minutes doing a planning exercise which will consist of steps 1 and 2 below. Once you have done your work there, proceed to analyze your circuit, and calculate the numerical value following the guidelines of step 3

1. **Placing the source:** Before starting, connect a source between the terminals where your input signal source should go. If you are working

a power transfer calculation, you will need both voltage and current of your source.

  (a) If the input signal is described only as a current or voltage at a node, then the other terminal by definition is ground.
  (b) If you plan to work using a reduction/transformation process, it may be easier if you use a source of the same type as your signal.
  (c) When setting up equations (nodal, loop, or other), you may use any type of source, but insure that the input signal type is included. In general, I prefer to include both voltage and current of the source in the equations, one of them to be considered "known". I do it this way because then I have access to all functions. If you don't do that, be sure to do the following:
    - If you use a current source but your input signal is a voltage, include the source voltage as a variable in the equations.
    - Inversely, for a voltage source but current signal, include the current of the source as a variable.
    - If you are to find power transfer, both voltage and current of the source *must* be included in the equations.

2. **Recognize your output signal** Identify your output signal, voltage or current. If it's power, associate it to a voltage, or current, or both. We refer next to voltage or output only.

  (a) If you are using a reduction/transformation method, try not to hide your output signal through the process.
    - If your method hides it, keep track of it by writing the equations or steps you will follow to recover it.
  (b) If you set up equations, like nodal, or loop, or a combination, be sure to associate your signal and your variables
    - When possible, choose the method that gives you the output signal directly; that is, a variable in your system of equations is the voltage or current that constitutes the output signal.
    - If your output signal is not one of the variables in the system of equations, be sure to write down an extra equation relating the variables and the output signal. This equation may be incorporated in the system, adding one variable –the output signal – or worked out separately.

3. **Find the numerical value of your function:**

  (a) If you used a source of different type as the signal, or else a source with a numeric value different to 1 (one), then divide the output voltage or current by the value of the input signal

8.2. FINDING THE NETWORK FUNCTIONS

(b) If your source has a numeric value of 1, then the numeric solution may be read directly

Table 8.1 summarizes how to find the numeric value of your function. The symbol $\doteq$ means "numerically equal". Half the values are obtained from the analysis result directly, the other half requires an additional step. Examples are given in the next section.

Table 8.1: Network Functions for unit value sources

Using a 1 A source:

Equivalent Resistance	$R_{eq} \doteq V_{in}$	Equivalent Conductance	$G_{eq} \doteq \dfrac{1}{V_{in}}$
Current transfer	$A_I \doteq I_{out}$	Voltage transfer	$A_v \doteq \dfrac{V_{out}}{V_{in}}$
Trans-resistance	$R_m \doteq V_{out}$	Trans conductance	$G_m \doteq \dfrac{I_{out}}{V_{in}}$

Using a 1 V source:

Equivalent Resistance	$R_{eq} \doteq \dfrac{1}{I_{in}}$	Equivalent Conductance	$G_{eq} \doteq I_{in}$
Current transfer	$A_I \doteq \dfrac{I_{out}}{I_{in}}$	Voltage transfer	$A_v \doteq V_{out}$
Transresistance	$R_m \doteq \dfrac{V_{out}}{I_{in}}$	Trans conductance	$G_m \doteq I_{out}$

As a final remark, notice that the functions are not completely independent. Some relations are the following:

$$P_v = A_v \times A_I = R_m \times G_m; \quad R_m = A_v \times R_{eq}; \text{ etc.}$$

However, only the port functions, equivalent resistance and conductance, are inverse one of the other, that is $R_{eq} \times G_{eq} = 1$. For the transfer functions, the inverse **is not** a transfer function. Be aware of that.

8.2.3 Further examples

Example 8.2 *For the circuit of Fig. 8.3, find the input resistance (that is, the equivalent resistance seen at the input) and the voltage gain (that is, A_v).*

Solution with current source: *Let us solve this problem, once using a current source, and the second time using a voltage source. For the current*

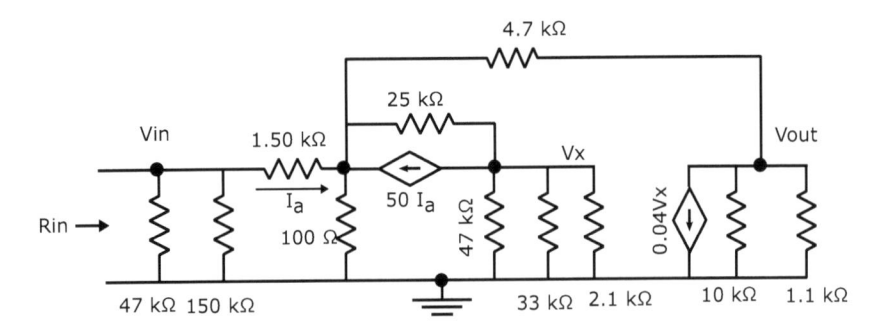

Figure 8.3: An example for Network Functions

source take Fig. 8.4. The input voltage is V_{N1}, or V_1 for short, and the output voltage is V_4.

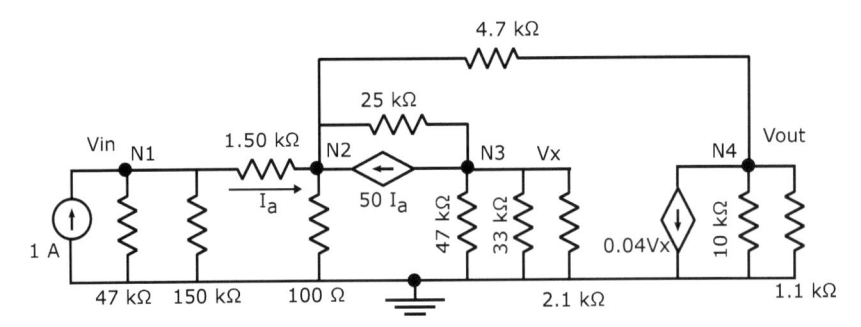

Figure 8.4: Solution using a current source

Using kΩ units for resistances, the nodal admittance matrix for this circuit, following the guidelines of chapter 6 may be obtained with

$$\mathbf{Y} = \begin{array}{c} \\ N1 \\ N2 \\ N3 \\ N4 \end{array} \begin{bmatrix} V_1 & V_2 & V_3 & V_4 \\ \frac{1}{47} + \frac{1}{150} + \frac{1}{1.5} & -\frac{1}{1.5} & 0 & 0 \\ -\frac{51}{1.5} & \frac{51}{1.5} + \frac{1}{0.1} + \frac{1}{25} + \frac{1}{4.7} & -\frac{1}{25} & -\frac{1}{4.7} \\ \frac{50}{1.5} & -\frac{50}{1.5} - \frac{1}{25} & \frac{1}{25} + \frac{1}{47} + \frac{1}{33} + \frac{1}{2.1} & 0 \\ 0 & -\frac{1}{4.7} & 40 & \frac{1}{4.7} + \frac{1}{1.1} + \frac{1}{10} \end{bmatrix}$$

(8.3)

Notice that the transconductance 0.04 S of the dependent source connected to the output was entered in mS units. That is 40 mS. The known vector is

8.2. FINDING THE NETWORK FUNCTIONS

$\mathbf{I_N} = [1, 0, 0, 0]^T$. *The solution* $\mathbf{Y}^{-1}\mathbf{I_N}$ *yields the following result, where I have labeled each row for easy reading:*

$$\begin{matrix} (Vin) \\ (V2) \\ (V3) \\ (Vout) \end{matrix} \begin{bmatrix} 22.854 \text{ E0} \\ 22.312 \text{ E0} \\ -30.252\text{E0} \\ 994.25\text{E0} \end{bmatrix} \qquad (8.4)$$

The Vin *value is numerically equal to the resistance seen by the current source, in* kΩ *units. Therefore* $R_{in} = 22.854$ kΩ. *The voltage transfer function* $A_v = V_{out}/Vin = 43.5$ *is obtained with*

`ans[4,1]/ans[1,1]` ENTER → 43.504

Notice also that $R_m = 994.25$ kΩ, *because this is the numerical value of* V_{out}. *The other functions cannot be determined because no load has been specified. We could think of an open circuit load, or else think of one of the resistances connected to ground and output as the load.*

Solution with voltage source: *Now let us introduce a voltage source as input, as illustrated in Fig. 8.5. Notice that we have included* I_{in} *as a variable to be used because this value is required to find the equivalent resistance seen by the source, i. e. , the input resistance.*

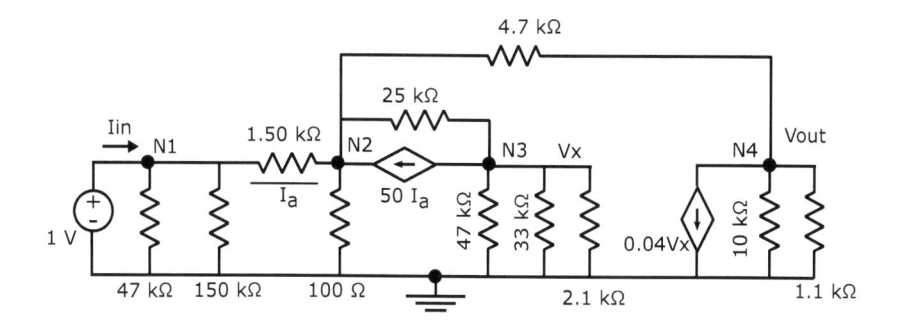

Figure 8.5: Solution using a voltage source

Again, following the guidelines and methods of chapter 6, the coefficient matrix and known vector are, respectively,

$$\mathbf{Y} = \begin{matrix} Node1 \\ Node2 \\ Node3 \\ Node4 \end{matrix} \begin{bmatrix} I_{in} & V_2 & V_3 & V_{out} \\ -1 & -\frac{1}{1.5} & 0 & 0 \\ 0 & \frac{51}{1.5} + \frac{1}{0.1} + \frac{1}{25} + \frac{1}{4.7} & -\frac{1}{25} & -\frac{1}{4.7} \\ 0 & -\frac{50}{1.5} - \frac{1}{25} & \frac{1}{25} + \frac{1}{47} + \frac{1}{33} + \frac{1}{2.1} & 0 \\ 0 & -\frac{1}{4.7} & 40 & \frac{1}{4.7} + \frac{1}{1.1} + \frac{1}{10} \end{bmatrix} \tag{8.5}$$

and

$$\mathbf{I_N} = \begin{matrix} Node1 \\ Node2 \\ Node3 \\ Node4 \end{matrix} \begin{bmatrix} V_{in} \\ -\left(\frac{1}{47} + \frac{1}{150} + \frac{1}{1.5}\right) \\ \frac{51}{1.5} \\ -\frac{50}{1.5} \\ 0 \end{bmatrix} \tag{8.6}$$

The solution for the equation, $\mathbf{Y}^{-1}\mathbf{I_N}$, results in

$$\begin{matrix} (Iin) \\ (V2) \\ (V3) \\ (Vout) \end{matrix} \begin{bmatrix} 43.755\text{E-}3 \\ 976.28\text{E-}3 \\ -1.3237\text{E}0 \\ 43.504\text{E}0 \end{bmatrix} \tag{8.7}$$

We see that $V_{out} = 43.504$ V, numerically equal to A_v, being the same result as before. Also, $I_{in} = 43.755$ E-3 mA, which is numerically equal to the equivalent conductance in mS units. Hence, in kΩ units the equivalent resistance is obtained as

ans[1,1]$^{-1}$ ENTER $\rightarrow$ 22.854E0, which is the same result as before.

8.3 Open and short circuit transfer functions

Sometimes it is necessary to calculate network functions under the special circumstances when the load is either an open circuit or a short circuit. Normally, the open circuit case is not so problematic, as example 8.2 has illustrated, since the output can be considered a voltage output with $I_{out} = 0$ A, which is precisely an open circuit.

8.3. OPEN AND SHORT CIRCUIT TRANSFER FUNCTIONS

Now, when the conditions require a short circuit, there are two forms to consider this problem. One, is to see if the short circuit current belongs to an element too when the short-circuit is introduced, in which case the problem is solved by considering this element to be the load.

The other, is when we require to introduce the short circuit and the current is not contained in an element. We may deal with the problem by applying the Kirchhoff Current equation there. Or else, we can introduce a voltage source at the desired terminals, and then the short circuit occurs when this source becomes 0.

Let us better illustrate this remark with an example.

Example 8.3 *Consider again the circuit of Fig. 8.3, but now we also want to find the output resistance R_{out}, which is the equivalent resistance seen from the output terminals.*

Since we are looking at the circuit from another pair of terminals, where we want to find an equivalent resistance, the problems becomes a candidate for superposition since, by definition, the equivalent resistance requires the subcircuit to be without independent sources.

*That means that the source signal must be 0 when calculating the output resistance. There is hence a difference whether we consider the signal to be a current or a voltage. In this case, **the source must be of the same type as the input signal!** The problem statement is incomplete, since it does not specify which type of input signal should we consider!*

Let us work it for a voltage input signal, and use the circuit of Fig. 8.5. This means in practical terms that R_{out} will be calculated with the input terminals in short circuit. I recommend the reader to use a current source for the open circuit case, and compare the results.

On the other hand, example 8.2 assumed always that the output was an open circuit. Therefore, to calculate the output resistance, we need to introduce a current source as our option. Otherwise, when applying superposition we will short circuit the output to ground! In conclusion, the circuit to be solved is that of Fig. 8.6.

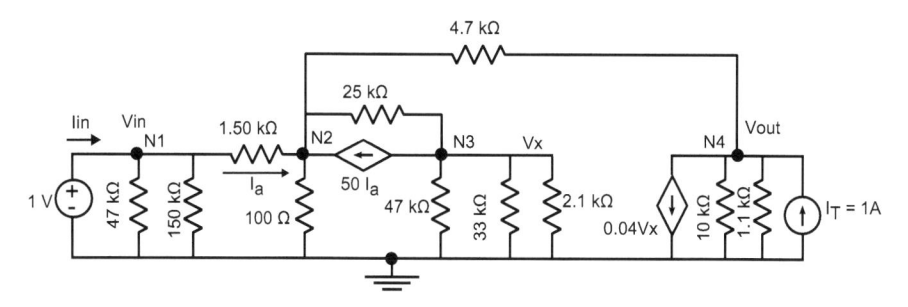

Figure 8.6: For an open circuit transfer function, and a short circuit output resistance calculation

For this circuit, the nodal equations used are in the form $\mathbf{YV} = \mathbf{I_N}$, where $\mathbf{Y}$ is the same as (8.5), while $\mathbf{I_N}$ is

$$\mathbf{I_N} = \begin{array}{c} Node1 \\ Node2 \\ Node3 \\ Node4 \end{array} \begin{bmatrix} V_{in} & I_T \\ -\left(\frac{1}{47} + \frac{1}{150} + \frac{1}{1.5}\right) & 0 \\ \frac{51}{1.5} & 0 \\ -\frac{50}{1.5} & 0 \\ 0 & 1 \end{bmatrix}$$

Observe that the first column is the vector (8.6). The solution for the system is

$$\begin{array}{c} (Iin) \\ (V2) \\ (V3) \\ (Vout) \end{array} \begin{bmatrix} 43.755\text{E-}3 & -257.24\text{E-}6 \\ 976.28\text{E-}3 & 385.86\text{E-}6 \\ -1.3237\text{E}0 & 22.681\text{E-}3 \\ 43.504\text{E}0 & 75.991\text{E-}3 \end{bmatrix} \quad (8.8)$$

The first column, which is what results if $I_T = 0$ A, has already been interpreted in the previous example. the second column corresponds to the case when $V_{in} = 0$ V. Interpreting results, the numerical value of V_{out} will be equal to the resistance seen by the current source, R_{out}, in kΩ units, when the input is short circuited. Therefore, $R_{out} = 76$ Ω.

In this example the reader should also appreciate the importance of interpretation when scaling has been applied, because of the units involved.

8.4 Closing remarks

There is an intimate relationship between the network functions and the principle of proportionality, and thus of homogeneity. The practical consequence is that finding the network function becomes a straightforward operation of analysis.

The signal concept is also important. Although it is irrelevant what type of signal source is used when all the functions to be calculated have the same input, the question of what kind of source is convenient becomes important when there is the need to calculate another function when excitation goes to another port, since then the original input source must be turned off, raising the question on whether we want a short circuit or an open circuit function. The reader can verify the different results obtained when sources are changed.

CHAPTER 9

Superposition: A powerful tool

The previous chapter applied the principle of proportionality to find the network functions of two terminal sub circuits without independent sources. But circuit theory and applications also include other important definitions and theorems such as Thevenin and Norton equivalents, two-port parameters, multiports, and so on.

This chapter presents superposition as the theoretical principle used to do these new calculations easily, while also also extending some results usually presented in textbooks.

9.1 Thevenin and Network Equivalents

We start by calculating Thevenin and Norton equivalent circuits. These are two important concepts in applied circuit theory. A first introduction was given before for particular cases.

9.1.1 Basic theory

Consider a linear resistive subcircuit N to which any kind of element may be connected, as shown in Fig. 9.1(a). Thevenin's theorem states that we can substitute the subcircuit by a voltage source in series with a resistance, as shown in (b), while Norton's theorem says that it can be substituted by a current source in parallel with a resistance, as shown in (c). These subcircuits are called, respectively, Thevenin and Norton equivalent circuits.

The voltage source V_{th} is called Thevenin's voltage. Since it is the voltage that we have in the particular case in which the element connected to the terminals is an open circuit, in which case $i = 0$, it is also called open circuit voltage V_{oc}. This is illustrated in Fig. 9.2(a).

Similarly, the current source I_N is called Norton's current. Since it is the current that occurs in the particular case in which the element connected to

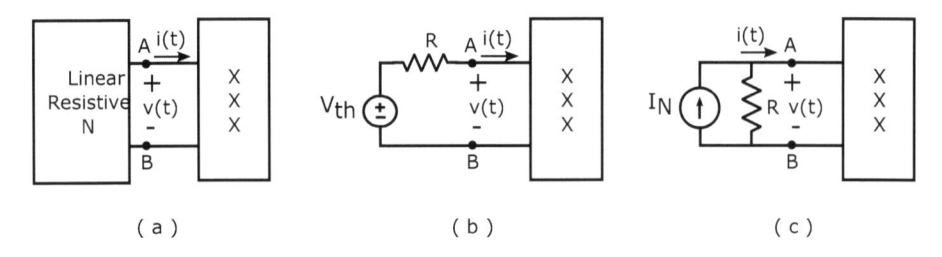

Figure 9.1: (a) General Configuration, (b) Thevenin's equivalent representation (c) Norton's Equivalent Representation

the terminals is a short circuit, in which case $v = 0$, it is also called short circuit current I_{sc}. This is illustrated in Fig. 9.2(b).

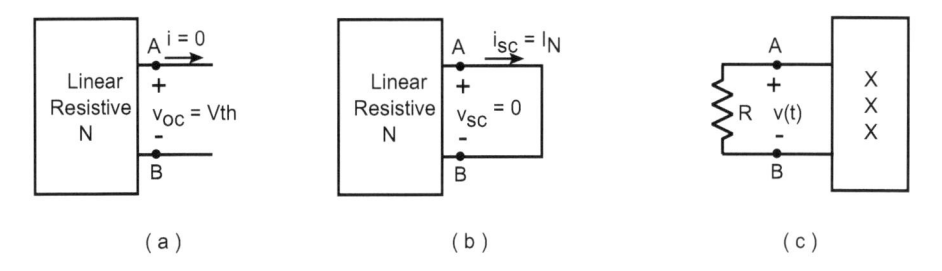

Figure 9.2: (a) Thevenin's voltage. (b) Norton's current (c) Equivalent Resistance R

The resistance R, called Thevenin's resistance or Norton's resistance, depending on the representation, is in fact the equivalent resistance seen by the element, which is calculated by turning off all independent sources in N, as illustrated in Fig. 9.2(c). Notice that Norton's circuit requires $R \neq 0 \, \Omega$ and Thevenin's representation requires $R \neq \infty$. When both representations exist they are in fact related by source transformation, that is,

$$V_{th} = RI_N \tag{9.1}$$

9.1.2 Finding equivalents: principles

Working the circuit with transformations, you can get an equivalent circuit directly by reducing the original subcircuit to the desired form. Hence, let us then establish the procedure convenient when working circuit equations.

At the terminals at which we want the representation, illustrated by Fig. 9.3(a), substitute the XXX element with a source, that can be either a current source as in Fig. 9.3(b) or a voltage source as in 9.3(c). *Observe the directions of the respective sources.*

9.1. THEVENIN AND NETWORK EQUIVALENTS

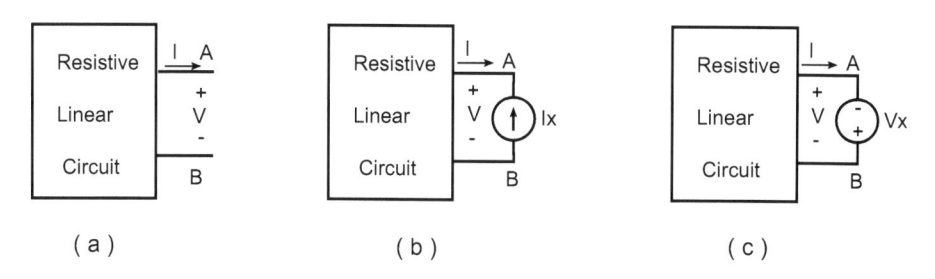

Figure 9.3: Finding Thevenin's and Norton's equivalent circuits: (a) Target subcircuit (b) using a current source or (b) using a voltage source

Consider the current source connection. By superposition, the voltage at the terminals is of the form $v = a + bI$. Here, a is the voltage when $Ix = 0$, that is, an open circuit, and $v = bIx$ when all other sources – internal to the sub circuit – are off. From the previous chapter, we know then that b is the equivalent resistance seen by the current source. In summary, the voltage v that results from the configuration in Fig. 9.3(b) is

$$v = V_{oc} + RI \tag{9.2}$$

Hence, the two parameters, Thevenin's Voltage and the equivalent resistance R, are directly read from the result if $Ix = 1$ A. Using (9.1) we may now find I_N.

Similarly, the current i in Fig. 9.3(c) is, by superposition, of the form $i = C + dV$, where C is the voltage when $V = 0$, and $i = dV$ when all other sources – internal to the sub circuit – are off. Therefore, taking into account the reference direction of i,

$$i = I_N + \frac{1}{R}V \tag{9.3}$$

Hence, in one step we have found Norton's current and the equivalent conductance $1/R$, whose inverse is R. With (9.1) we may now find V_{th}.

9.1.3 Finding equivalents: examples

Let us illustrate principles with examples. Let us work the first one using both methods.

Example 9.1 *We are interested in finding the Norton and Thevenin equivalents for the subcircuit of Fig. 9.4 as seen from terminals A and B.*

This will be done using a current source, and a voltage source, separately. The arrangements for the respective cases are shown in Fig. 9.5. The current source case is prepared for nodal analysis in (a), while that with the voltage source for loop analysis in (b). A source transformation was included for the sake of easier procedure in the second arrangement.

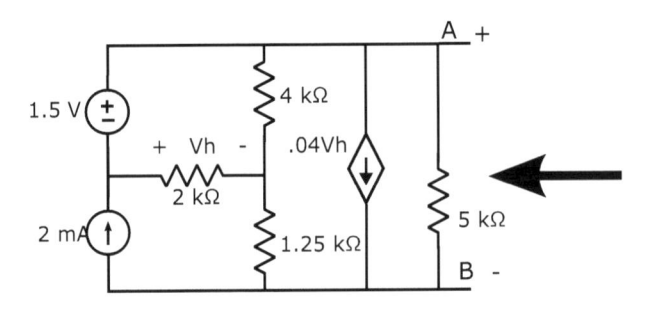

Figure 9.4: Target subcircuit for Thevenin and Norton equivalent circuits

(a)

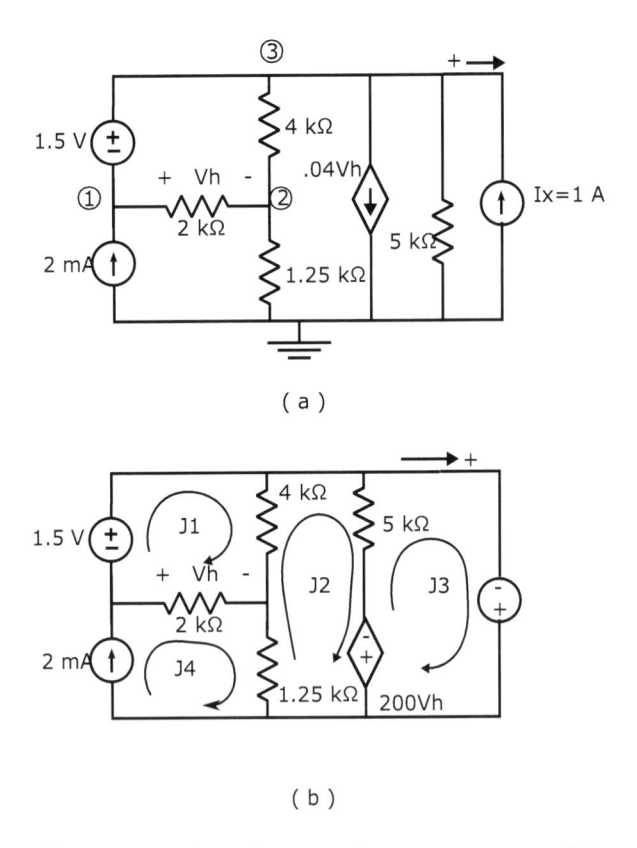

(b)

Figure 9.5: (a) Prepared for solving with current source; (b) prepared for voltage source solution

9.1. THEVENIN AND NETWORK EQUIVALENTS

Solving with the current source: *Since we are not interested on the current flowing in the voltage source, let us use the "super node" equations, joining nodes 1 and 3 in one equation, as shown in (9.4) below. If you have problems inserting the "0.04", remember that the dependent source is going into node 3, which belongs to the supernode, and thus is negative. Write down the equations by hand first if you have doubts.*

$$\begin{array}{c} \\ N(1,3) \\ N2 \\ 1.5V \end{array} \begin{bmatrix} \overset{V_1}{0.04 + \frac{1}{2E3}} & \overset{V_2}{-0.04 - \frac{1}{2E3} - \frac{1}{4E3}} & \overset{V_3}{\frac{1}{4E3} + \frac{1}{5E3}} \\ -\frac{1}{2E3} & \frac{1}{2E3} + \frac{1}{4E3} + \frac{1}{1250} & -\frac{1}{4E3} \\ -1 & 0 & 1 \end{bmatrix} \begin{bmatrix} v_1 \\ v_2 \\ v_3 \end{bmatrix} = \begin{bmatrix} \overset{"In"}{0.002} & \overset{Ix}{1} \\ 0 & 0 \\ 1.5 & 0 \end{bmatrix}$$

(9.4)

The solution for this system is, after entering the coefficient matrix as yn, *and the knowns as* in,

$$\text{simult(yn,in)} \rightarrow \begin{bmatrix} 526.74\text{E-3} & 47.098\text{E0} \\ 496.818\text{E-3} & 22.789\text{E0} \\ 2.027\text{E0} & 47.098\text{E0} \end{bmatrix}$$

Since we are looking the subsystem between node 3 and ground, the solution comes out from reading the result for this voltage, at the third row: Vth = 2.027 V *and an equivalent resistance* Rth = 47.098 Ω. *The Norton Current source is found using (9.1), working with the values already in the calculator:*

ans[3,1]/ans[3,2] → 42.032E-3

which we interpret as a short circuit Norton current IN = 42.032 mA.

Solving with the voltage source: *Let us now take the circuit in Fig. 9.5(b). Again, since we are not interested on the voltage at the 2 mA current source, we can write the system as* $\mathbf{Z_M} \mathbf{J_M} = \mathbf{V_L}$, *where* $\mathbf{J_M} = \begin{bmatrix} J_1, J_2, J_3 \end{bmatrix}^T$,

$$\mathbf{Z_M} = \begin{array}{c} Loop1 \\ Loop2 \\ Loop3 \end{array} \begin{bmatrix} \overset{J_1}{4E3 + 2E3} & \overset{J_2}{-4E3} & \overset{J_3}{0} \\ -4E3 + 200(2E3) & 4E3 + 1250 + 5E3 & -5E3 \\ -200(2E3) & -5E3 & 5E3 \end{bmatrix} \quad (9.5a)$$

and

$$\mathbf{V_L} = \begin{matrix} Loop1 \\ Loop2 \\ Loop3 \end{matrix} \begin{bmatrix} \overset{VL}{0.002(2E3) + 1.5} & \overset{Vx}{0} \\ 0.002(1250) + 0.002(200)(2E3) & 0 \\ -0.002(200)(2E3) & 1 \end{bmatrix} \quad (9.5b)$$

The solution for this system is, after entering the coefficient matrix as `zm`, and the knowns as `vl`,

```
         [ 2.508E-3    258.065E-6 ]
zm^-1 * vl →  2.387E-3    387.097E-6
         [ 43.032E-3   21.232E-3  ]
```

Let us store this answer in variable B. Looking at the solution for J3, the first column shows Norton's current IN = 43.032 mA, which is the same as before. The equivalent conductance is 21.232E-3 S whose inverse is the equivalent resistance:

`1/B[3,2]` $\to$ 47.098E0, which corresponds to Rth = 47.098 Ω

The Thevenin's voltage is obtained with the relationship $V_{th} = I_N R_{th} = I_N/G_{eq}$:

`B[3,1]/B[3,2]` $\to$ 2.027E0, which coincides with the previous result Vth = 2.027 V

The example shows that the analysis method or preferred source does not determine the validity of the process. It must be emphasized that we are assuming that the system has a solution, since there are situations in which this does not happen. For example, in those cases in which a voltage source is in parallel with the port (R=0) or a current source in series with the port (R = ∞).

Exercise suggested: to solve the problem now by generating the coefficient matrices with the programs for nodal and mesh analyses introduced in previous chapters.

When the circuit contains symbolic sources, then the Thevenin Voltage will also be of symbolic type. It can also be obtained as a function of several sources. The following example illustrates this.

Example 9.2 *Assume that the independent sources in example 9.1 are symbolic. That is, in Fig. 9.4 assume that instead of the 1.5 V and 2 mA sources we have E_a and I_b, respectively. We can find the equivalent representations using the methods that have been used for dealing with symbolic sources. Let us work it next.*

9.1. THEVENIN AND NETWORK EQUIVALENTS

Solving with the current source: The coefficient matrix in 9.4 will not change, but the "known" matrix I_N becomes

$$\mathbf{I_N} = \begin{matrix} Node(1,3) \\ Node2 \\ V-src \end{matrix} \begin{matrix} Va & Ib & Ix \\ \begin{bmatrix} 0 & 1 & 1 \\ 0 & 0 & 0 \\ 1 & 0 & 0 \end{bmatrix} \end{matrix} \quad (9.6)$$

The solution for this system is, after entering again the coefficient matrix as `yn` –which you might already have –, and the knowns as `in`,

$$\mathtt{yn^{-1} * in} \rightarrow \begin{bmatrix} 288.4\text{E-}3 & 47.1\text{E}0 & 47.098\text{E}0 \\ 300.8\text{E-}3 & 22.8\text{E}0 & 22.789\text{E}0 \\ 1.288\text{E}0 & 47.1\text{E}0 & 47.098\text{E}0 \end{bmatrix}$$

Looking at the third row, we interpret the open circuit voltage as a linear combination of the sources: Vth = 1.288 Va + 47.1 Ib. The equivalent resistance Rth = 47.1 Ω appears here as expected. The Norton Current source is found using (9.1), working with the values already in the calculator:

`ans[3]/ans[3,3]` → [27.355E-3 1.E0 1.E0]

which we interpret as IN = 27.355E-3 Va + Ib. The equivalent representations for these results are shown in Fig. 9.6.

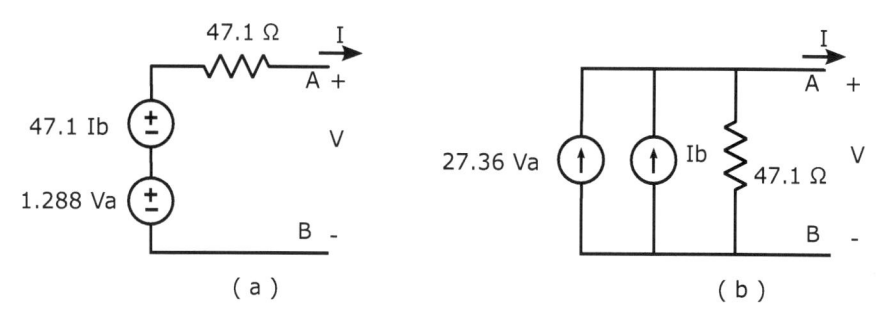

Figure 9.6: (a) Thevenin Equivalent for this example; (b) Norton Equivalent for this example

As you might have imagined from this example, it is possible to extend the applications of superposition to calculate another variations such as for example when one of the sources is dependent on a magnitude which is not part of the subcircuit itself. We leave that discussion outside from this book.

Further applications of the Thevenin and Norton's theorems include among others the maximum power transfer theorem, extensions to n-ports, and so on. It will be used again here applications that will be left for Chapter 11. Let us for the moment look at another topic.

9.2 Calculating two-port parameters

The topic of two-port is important enough to deserve its own chapter. Here I limit myself to present the application of superposition to find the parameters. I introduce also an extension of the concept for extended Thevenin and Norton equivalents.

If the reader is not familiar with two-ports yet, a brief introduction is found in the next chapter. Also, the reader may consult circuit textbooks.

9.3 Finding two-port parameters from equations

Two-port parameters may be calculated in different ways:

a) By direct application of definitions: Once we short-circuit or open-circuit terminals, we apply the methods studied in previous chapters. This is the usual way in which they are taught in textbooks. And important also.

b) Obtaining another set of parameters and using conversion formulas. Tables of translation formulas are usually available from textbooks, and also discussed in the next chapter.

c) Connecting sources to both ports and solving the equations for dependent port variables in terms of the independent variables. This is the method to be discussed next. The parameters result from coefficient comparison of results

I illustrate the method using one circuit and deriving two sets of parameters. The equations must include both current and voltages of the added sources.

Example 9.3 *Find the h parameters of the two port shown in Fig. 9.7.*

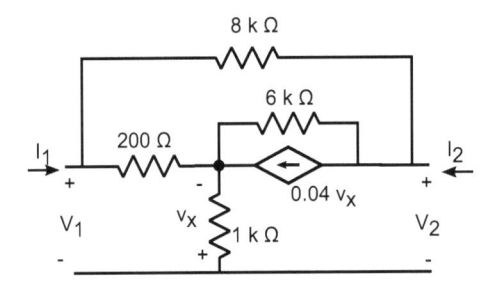

Figure 9.7: Two port example

I will work this example using mesh equations, with a previous source transformation, as shown in Fig. 9.8. Notice that I do not show the type

9.3. FINDING TWO-PORT PARAMETERS FROM EQUATIONS

of source being used. This is irrelevant for our purposes since the method is based on variable manipulation.

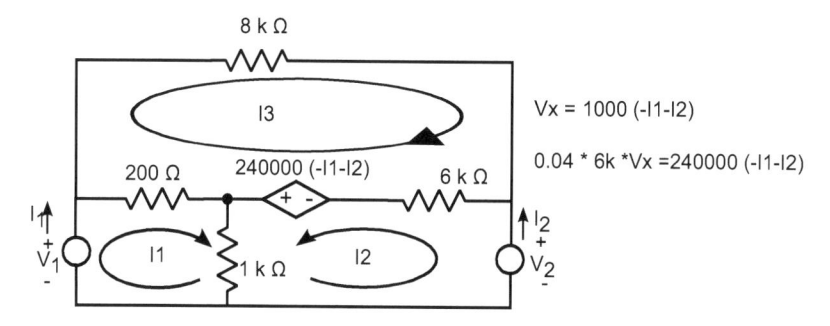

Figure 9.8: Two-port example arranged for loop equations

Before we start working, let us consider what we are looking for, in order to see how to write down the equations. Since we want h-parameters, that means that our goal is to arrive at equations of the form

$$V_1 = h_{11} I_1 + h_{12} V_2$$
$$I_2 = h_{21} I_1 + h_{22} V_2$$

Hence, V_1 and I_2 are supposed to be unknown, together with I_3, while I_1 and V_2 are considered "symbolic" known sources. Using the methods for loop analysis that we have already studied for such a case, **not forgetting to include the voltage at port 1 as variable,** *we arrive at*

$$\begin{array}{c} \\ Loop1 \\ Loop2 \\ Loop3 \end{array} \begin{array}{ccc} V_1 & I_2 & I_3 \\ \begin{bmatrix} -1 & 1000 & -200 \\ 0 & 247000 & 6000 \\ 0 & 246000 & 14200 \end{bmatrix} \end{array} \begin{bmatrix} V_1 \\ I_2 \\ I_3 \end{bmatrix} = \begin{array}{cc} I_1 & V_2 \\ \begin{bmatrix} -1200 & 0 \\ -241000 & 1 \\ -239800 & 0 \end{bmatrix} \end{array} \quad (9.7)$$

Storing matrices in **p**, **q**, *the solution* `simult(p,q)` *for this system yields*

$$\begin{bmatrix} V_1 \\ I_2 \\ I_3 \end{bmatrix} = \begin{array}{cc} I_1 & V_2 \\ \begin{bmatrix} 218.1747 & 00.03121 \\ -0.9764 & 7 \times 10^{-6} \\ 0.02727 & -0.000121 \end{bmatrix} \end{array}$$

The H-matrix is formed with the rows corresponding to V_1 and I_2. In this case, the first two rows in that order. Hence, with `h:= submat(ans,1,1,2,2)` *we get*

$$\mathbf{H} = \begin{bmatrix} 218.17 & 31.21E - 3 \\ -976.37E - 3 & 6.99E - 6 \end{bmatrix}$$

Now, for the same circuit let us assume that we are interested in the B parameters. That is, our goal is to arrive at equations of the form

$$V_2 = b_{11} V_1 - b_{12} I_1$$
$$I_2 = b_{21} V_1 - b_{22} I_1$$

Hence, V_2 and I_2 are supposed to be unknown, while V_1 and I_1 are considered known. Observe the sign for the term of current I_1. This can be accomplished by changing the sign for the column of I_1. This is shown in the next expression with the label $-I_1$. The reader can verify now that the equations are

$$\begin{array}{c} Loop1 \\ Loop2 \\ Loop3 \end{array} \begin{bmatrix} \overset{V_2}{0} & \overset{I_2}{1000} & \overset{I_3}{-200} \\ -1 & 247000 & 6000 \\ 0 & 246000 & 14200 \end{bmatrix} \begin{bmatrix} V_2 \\ I_2 \\ I_3 \end{bmatrix} = \begin{bmatrix} \overset{V_1}{1} & \overset{``-I_1''}{1200} \\ 0 & 241000 \\ 0 & 239800 \end{bmatrix} \quad (9.8)$$

The solution for this system is

$$\begin{bmatrix} V_2 \\ I_2 \\ I_3 \end{bmatrix} = \begin{bmatrix} \overset{V_1}{32.041E0} & \overset{-I_1}{6.99E3} \\ 223.97E-6 & 1.025E0 \\ -3.88E-3 & -873.82E-3 \end{bmatrix}$$

Matrix B is now given by the first two rows of the answer, that is

$$\mathbf{B} = \begin{bmatrix} 32.041E0 & 6.99E3 \\ 223.97E-6 & 1.025E0 \end{bmatrix}$$

The previous example has shown the method for calculation directly from equations. Now, what do you want your programmable calculator for if you are always going to write the equations by yourself? Nah! Let us take advantage of the programs we have written. This become specially true if the circuit is big enough to make you feel insecure.

Using the nodal and mesh programs

When you write the nodal equations with current signals in mind, then the solution for the **Z** parameters is almost direct. Consider the structure of Fig. 9.10. If both ports share a terminal, as in inset (b), then potentials V_1 and V_2 become the port voltages and the solution is almost direct, as shown below in example 9.4. If, the ports do not share terminals, then select one of the port terminals as ground, as shown in (c). Then potential V_1 is one of the port voltages while $V_2 - V_3$ is the other. We can either include an additional equation or else build up the **Z** matrix after solving. The examples below

9.3. FINDING TWO-PORT PARAMETERS FROM EQUATIONS 185

will illustrate these cases. For the matrices in the figure, remember that $\doteq$ means "numerically equal".

$$Z \doteq \begin{bmatrix} Vn[1] \\ Vn[2] \end{bmatrix}$$

(a)   (b)

$$Z \doteq \begin{bmatrix} Vn[1] \\ Vn[2]-Vn[3] \end{bmatrix}$$

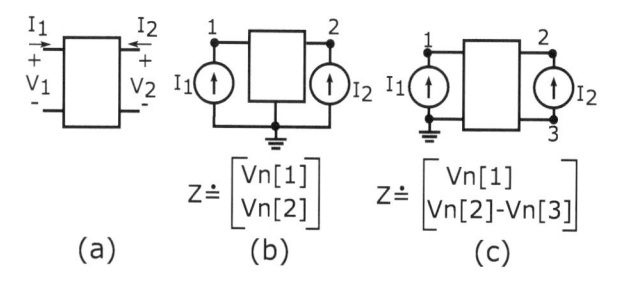

(c)

Figure 9.9: To find Z Parameters with nodal analysis: (a) Two-port definition; (b)Three-terminal case; (c) Ports without common terminal

(a)

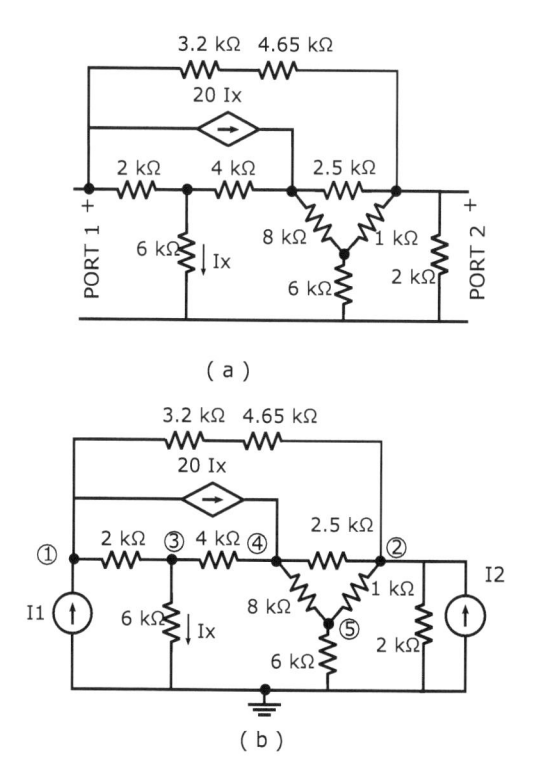

(b)

Figure 9.10: Two-port example using nodal program: (a) Two-port definition; (b) Prepared for analysis

Example 9.4 *Consider the two-port of Fig. 9.10(a). We want to obtain the* **Z** *parameters for this two-port, using the program we have developed for*

nodal analysis. Inset (b) shows the port prepared for nodal analysis. The description of the circuit, without the independent sources, is

$$\mathbf{R} = \begin{bmatrix} 2000 & 1 & 3 \\ 3200 + 4650 & 1 & 2 \\ 6000 & 3 & 6 \\ 4000 & 3 & 4 \\ 8000 & 4 & 5 \\ 6000 & 5 & 6 \\ 2500 & 4 & 2 \\ 1000 & 2 & 5 \\ 2000 & 2 & 6 \end{bmatrix} \quad G = \begin{bmatrix} 20/6E3 & 1 & 4 & 3 & 6 \end{bmatrix} \quad (9.9)$$

Now, use the program with the input `nodal1(5,r,0,g)` [enter]. You can either type the name or call the program with the [var] key. We have introduced a 0 value for the sources because we are interested only on the nodal admittance matrix, which is

yn =
$$\begin{bmatrix} 627.4E-6 & -127.4E-6 & 2.833E-3 & 0.0 & 0.0 \\ -127.4E-6 & 2.027E-3 & 0.0 & -400E-6 & -1.E-3 \\ -500.E-6 & 0.0 & 916.67E-6 & -250.E-6 & 0.0 \\ 0.0 & -400E-6 & -3.583E-3 & 775.E-6 & -125E-6 \\ 0.0 & -1.E-3 & 0.0 & -125.E-6 & 1.292E-3 \end{bmatrix}$$
(9.10)

With this result we are ready to introduce the sources. Remember they are symbolic:

$$\mathbf{B} = \begin{bmatrix} 1 & 0 \\ 0 & 1 \\ 0 & 0 \\ 0 & 0 \\ 0 & 0 \end{bmatrix}$$

The solution for `simult(yn,b)` is

$$\begin{bmatrix} -734.81 & -586.16 \\ 1.345E3 & 1.489E3 \\ 576.128 & 196.744 \\ 3.582E3 & 1.894E3 \\ 1.388E3 & 1.336E3 \end{bmatrix}$$

From this result, we obtain the **Z** parameters with any of the following two options:

9.3. FINDING TWO-PORT PARAMETERS FROM EQUATIONS 187

subMat(ans,1,1,2,2) *or* augment(ans[1];ans[2])

For the first option hit menu *7 1 7, while for the second one* menu *7 1 8. The final result is*

$$\mathbf{Z} = \begin{bmatrix} -734.81 & -586.16 \\ 1.345E3 & 1.489E3 \end{bmatrix}$$

Let us illustrate another situation, this time with the ports not sharing a common node.

Example 9.5 *Consider the the same circuit of the previous example, but now the two ports are as shown in of Fig. 9.11(a). We want to obtain the* **Z** *parameters for this two-port, using the program we have developed for nodal analyisis. Inset (b) shows the port prepared for nodal analysis.*

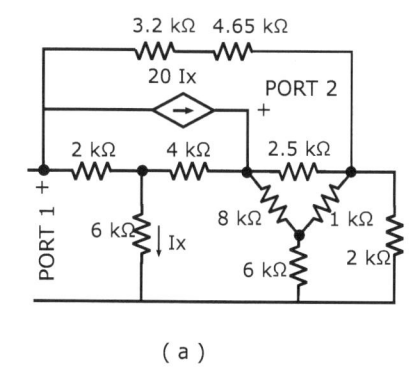

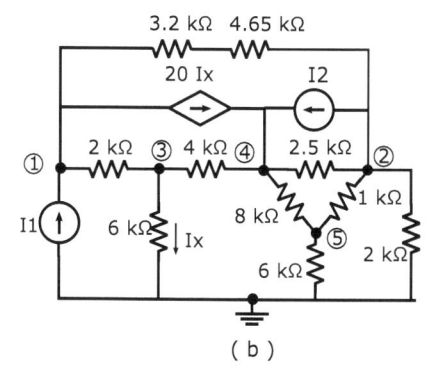

Figure 9.11: Another two-port example using nodal program: (a) Two-port definition; (b) Prepared for analysis

Since the circuit is the same as before, we can use the same description (9.9) for the circuit and obtain the same nodal admittance matrix (9.10).

However, now the port voltages are:

Port 1: Potential V_1;   Port 1: Potential V_4 - Potential V_2

The source matrix is,

$$B = \begin{bmatrix} 1 & 0 \\ 0 & -1 \\ 0 & 0 \\ 0 & 1 \\ 0 & 0 \end{bmatrix}$$

The solution to `simult(yn,b)` *is now*

$$\begin{bmatrix} -734.81 & -704.11 \\ 1.345E3 & -90.614 \\ 576.128 & 151.84 \\ 3.582E3 & 1.965E3 \\ 1.388E3 & 120.00 \end{bmatrix}$$

Observe that the first column is identical to that of the previous example. This result was expected because when I2=0 we have the same circuit for both cases. From this result, we can obtain the **Z** *parameters with*

`augment(ans[1];ans[4]-ans[2])` enter

$$Z = \begin{bmatrix} -734.81 & -704.11 \\ 2.237E3 & 2.056E3 \end{bmatrix}$$

With the previous examples you see that it is possible to obtain two-port parameters with superposition. The first example provided a way to do so by manipulation of equations so the desired independent variables are treated as "known" data. On the other hand, the latter examples used the nodal program to obtain Z-parameters.

Using the mesh analysis program, or loop analysis in general, it is possible to obtain Y parameters. I leave to the reader to look for examples.

Now, if the circuit is complex and you need other parameters than the Z or Y, you can always manipulate the equations after getting the nodal admittance matrix. Yet, it is better in my opinion to work parameter transformations as explained in the following chapter. In the particular case of Z and Y parameters, $\mathbf{Z}^{-1} = \mathbf{Y}$. This means that the nodal analyisis program becomes handy for those circuits which are not planar.

9.3.1 "Active two-ports"

The theory for two ports assumes that there is no independent source in the subcircuit. The same principle that applies to equivalent resistance. Now,

9.3. FINDING TWO-PORT PARAMETERS FROM EQUATIONS

what happens if there are independent sources in the subcircuit? We can proceed in a similar way as we did by considering Thevenin's and Norton's equivalents: add a source in series or parallel, depending on the case.

Without discussing the theory in depth, let us take an example, working with the H-parameters. The example uses one independent source but, as you might imagine, is applicable to circuits with more sources as well. Including symbolic or time dependen ones, which you already know how to deal with using superposition.

Example 9.6 *In the two port of Fig. 9.7 on page 182 tinclude a 6 V source in series with the 1 kΩ resistance. Prepare the system for analysis as shown in Fig. 9.12. The objective is to get an expression for V_1 and I_2 in terms of I_1 and V_2, as well as the voltage source. In other words, to obtain the h-parameters without turning off the source, and combine with the effect of the source.*

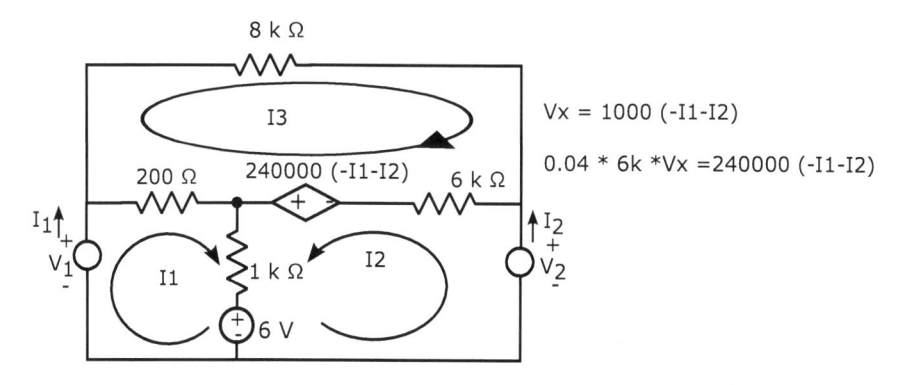

Figure 9.12: Two-port example with an independent soruce arranged for loop equations

We begin by setting up the equations in the same way we did it in example 9.3, but now including the 6 V source. The reader can verify that the set of equations in matrix form is now as follows. Labels have been included for easy reference. You may store the respective matrices in **A** and **B**.

$$\begin{array}{c} \\ Loop1 \\ Loop2 \\ Loop3 \end{array} \begin{array}{ccc} V_1 & I_2 & I_3 \\ \left[\begin{array}{ccc} -1 & 1000 & -200 \\ 0 & 247000 & 6000 \\ 0 & 246000 & 14200 \end{array}\right] \end{array} \begin{bmatrix} V_1 \\ I_2 \\ I_3 \end{bmatrix} = \begin{array}{c} \\ \\ \\ \end{array} \begin{array}{ccc} 6V & I_1 & V_2 \\ \left[\begin{array}{ccc} -6 & -1200 & 0 \\ -6 & -241000 & 1 \\ 0 & -239800 & 0 \end{array}\right] \end{array} \quad (9.11)$$

The solution, with `simult(A,B)` or $\mathbf{A}^{-1}\mathbf{B}$ yields the following. I have used engineering notation for easiness.

$$\begin{bmatrix} V_1 \\ I_2 \\ I_3 \end{bmatrix} = \begin{bmatrix} \overset{6V}{5.8127E0} & \overset{I_1}{218.17E0} & \overset{V_2}{31.21E-3} \\ -41.942E-6 & -976.37E-3 & 6.99E-6 \\ 726.56E-6 & 27.272E-3 & -121.1E-6 \end{bmatrix}$$

With ha:= submat(ans,1,1,3,2) *we generate the "active" h-parameter matrix:*

$$\mathbf{Ha} = \begin{bmatrix} 5.8127E0 & 218.17E0 & 31.21E-3 \\ -41.942E-6 & -976.37E-3 & 6.99E-6 \end{bmatrix}$$

where we can identify the right columns as the original one obtained in example 9.3. Thus, at port 1 we see a Thevenin's type circuit representation, while at por 2 we see a Norton's type one. I invite the reader to draw them. As a hint, you may look at Fig. 10.2 on page 200.

9.4 Closing remarks

This chapter has shown two applications of superposition using matrices. The mechanism follows in the different applications the same basic principles. Namely, write down the equations in a convenient form so that the solution for the problem you're trying to solve comes out directly, via of course the proper interpretation.

The application of the superposition theorem was applied just to two sort of problems. First, to find the Thevenin and Norton equivalent representations. And second, to calculate the parameters of two-port sub circuits. It should be obvious by now that there are many more possible cases where this method is handy.

CHAPTER 10

Two-port and three-terminal networks

Chapter 9 included a section on how to find two port parameters using the property of superposition. In this chapter we focus on some theory and applications of two-port networks, leaving more in depth study for the reader to follow. An interesting feature is that two-ports are attractive for quick calculations, because most applications are mainly based on formulas that can be programmed.

10.1 Basic Definitions

Any device or sub circuit with two or more terminals available for connection is a *multiterminal network*, or *multipole*. If one terminal is taken as a datum or ground with respect to which the potentials of other N terminals are referenced, it is called (N+1)-terminal network, or (N+1)-pole. When the reference node is not connected to any element of the network, then it is a *floating* N-pole or floating multipole.

On the other hand, two terminals form a *port* if the current entering one terminal is always equal to the current leaving the other. A 2N terminal network is an *N-port network*, or simply N-port, when the terminals can be associated by pairs to form ports.

Fig. 10.1 shows the general representation of two-ports and three terminal sub circuits. A three pole may be considered a special case of two-port with one pole common to both ports. The difference between both structures may be important when it comes to applications, but this topic falls beyond our scope. We focus here only in common properties. Hereafter, both will be referred as "two-ports" unless distinction is needed.

Ports 1 and 2 are often called *input port* and *output port*, respectively, and this convention is used in naming parameters. The reader should however be aware that these names have real meaning only in specific applications.

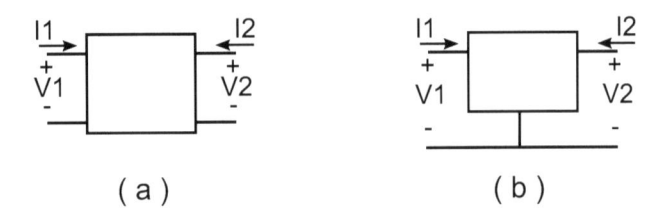

Figure 10.1: (a) Two port and (b) three-terminal networks

10.2 Two port parameters

Two linear equations characterize a two-port. With four variables involved, V1, V2, I1 and I2, two of them must be selected as independent variables. Six different choices are possible yielding different parameters to characterize the two-port. Some parameters may not exist for particular circuits.

The definition of the parameters assumes that the two-port circuit does not contain any independent source. To find the parameters, therefore, it could be necessary to first turn those sources off. Example 9.6 has illustrated that in fact turning off the sources is not absolutely necessary if you work with superposition.

We describe next the different representations.

10.2.1 Definition of Parameters

The six sets of two-port parameters are defined in table 10.1 on the next page. The notations OC, for open-circuit, and SC for short-circuit, arise from the conditions set for the parameter definition. When a current is zero, it becomes an open circuit, and a short-circuit when it is a zero voltage.

Notice the negative sign for the independent current in the chain parameters. This negative sign stems from a change in the reference direction for the respective current. We shall see the advantage later.

The chain parameters are also known as ABDC parameters because originally the equations were written, in matrix form, as

$$\begin{bmatrix} V_1 \\ I_1 \end{bmatrix} = \begin{bmatrix} A & B \\ C & D \end{bmatrix} \begin{bmatrix} V_2 \\ -I_2 \end{bmatrix}$$

Similarly, for the B-parameters, the name used was A'B'C'D' for the notation

10.2. TWO PORT PARAMETERS

Table 10.1: Two Port Parameters

Open Circuit impedances or Z-parameters

Matrix equation:
$$\begin{bmatrix} V_1 \\ V_2 \end{bmatrix} = \begin{bmatrix} z_{11} & z_{12} \\ z_{21} & z_{22} \end{bmatrix} \begin{bmatrix} I_1 \\ I_2 \end{bmatrix}$$

OC. Input impedance:
$$z_{11} = \left.\frac{V_1}{I_1}\right|_{I_2=0}$$
OC. reverse transimpedance:
$$z_{12} = \left.\frac{V_1}{I_2}\right|_{I_1=0}$$
OC. forward transimpedance:
$$z_{21} = \left.\frac{V_2}{I_1}\right|_{I_2=0}$$
OC. output impedance:
$$z_{22} = \left.\frac{V_2}{I_2}\right|_{I_1=0}$$

Hybrid H-Parameters

Matrix equation:
$$\begin{bmatrix} V_1 \\ I_2 \end{bmatrix} = \begin{bmatrix} h_{11} & h_{12} \\ h_{21} & h_{22} \end{bmatrix} \begin{bmatrix} I_1 \\ V_2 \end{bmatrix}$$

SC input impedance:
$$h_{11} = \left.\frac{V_1}{I_1}\right|_{V_2=0}$$
OC reverse voltage gain:
$$h_{12} = \left.\frac{V_1}{V_2}\right|_{I_1=0}$$
SC current gain:
$$h_{21} = \left.\frac{I_2}{I_1}\right|_{V_2=0}$$
OC admittance:
$$h_{22} = \left.\frac{I_2}{V_2}\right|_{I_1=0}$$

Chain A parameters or ABCD parameters

Matrix equation:
$$\begin{bmatrix} V_1 \\ I_1 \end{bmatrix} = \begin{bmatrix} a_{11} & a_{12} \\ a_{21} & a_{22} \end{bmatrix} \begin{bmatrix} V_2 \\ -I_2 \end{bmatrix}$$

OC voltage gain:
$$\frac{1}{a_{11}} = \left.\frac{V_2}{V_1}\right|_{I_2=0}$$
SC transadmittance:
$$\frac{1}{a_{12}} = \left.\frac{-I_2}{V_1}\right|_{V_2=0}$$
OC transimpedance:
$$\frac{1}{a_{21}} = \left.\frac{V_2}{I_1}\right|_{I_2=0}$$
SC current gain:
$$\frac{1}{a_{22}} = \left.\frac{-I_2}{I_1}\right|_{V_2=0}$$

Short Circuit admittances or Y-parameters

Matrix equation:
$$\begin{bmatrix} I_1 \\ I_2 \end{bmatrix} = \begin{bmatrix} y_{11} & y_{12} \\ y_{21} & y_{22} \end{bmatrix} \begin{bmatrix} V_1 \\ V_2 \end{bmatrix}$$

SC. Input admittance:
$$y_{11} = \left.\frac{I_1}{V_1}\right|_{V_2=0}$$
SC. Reverse transadmittance:
$$y_{12} = \left.\frac{I_1}{V_2}\right|_{V_1=0}$$
SC. forward transadmittance:
$$y_{21} = \left.\frac{I_2}{V_1}\right|_{V_2=0}$$
SC. output admittance:
$$y_{22} = \left.\frac{I_2}{V_2}\right|_{V_1=0}$$

Inverse hybrid G-Parameters

Matrix equation:
$$\begin{bmatrix} I_1 \\ V_2 \end{bmatrix} = \begin{bmatrix} g_{11} & g_{12} \\ g_{21} & g_{22} \end{bmatrix} \begin{bmatrix} V_1 \\ I_2 \end{bmatrix}$$

OC input admittance:
$$g_{11} = \left.\frac{I_1}{V_1}\right|_{I_2=0}$$
SC reverse current gain:
$$g_{12} = \left.\frac{I_1}{I_2}\right|_{V_1=0}$$
OC voltage gain:
$$g_{21} = \left.\frac{V_2}{V_1}\right|_{I_2=0}$$
SC output impedance:
$$g_{22} = \left.\frac{V_2}{I_2}\right|_{V_1=0}$$

Inverse chain B parameters or A'B'C'D' parameters

Matrix equation:
$$\begin{bmatrix} V_2 \\ I_2 \end{bmatrix} = \begin{bmatrix} b_{11} & b_{12} \\ b_{21} & b_{22} \end{bmatrix} \begin{bmatrix} V_1 \\ -I_1 \end{bmatrix}$$

OC reverse voltage gain:
$$\frac{1}{b_{11}} = \left.\frac{V_1}{V_2}\right|_{I_1=0}$$
SC reverse transadmittance:
$$\frac{1}{b_{12}} = \left.\frac{-I_1}{V_2}\right|_{V_1=0}$$
OC reverse transimpedance:
$$\frac{1}{b_{21}} = \left.\frac{V_1}{I_2}\right|_{I_1=0}$$
SC reverse current gain:
$$\frac{1}{b_{22}} = \left.\frac{-I_1}{I_2}\right|_{V_1=0}$$

194 CHAPTER 10. TWO-PORT AND THREE-TERMINAL NETWORKS

$$\begin{bmatrix} V_2 \\ I_2 \end{bmatrix} = \begin{bmatrix} A' & B' \\ C' & D' \end{bmatrix} \begin{bmatrix} V_1 \\ -I_1 \end{bmatrix}$$

10.2.2 Transforming parameters

With the exception of chain parameters, there may be sets of parameters that do not exist for one particular two-port network. This section works parameter conversion assuming the target set exists. Two main algorithms can be considered: work conversion as a problem of equations solving, or applying transformation formulas. These can be used to build up your set of user defined functions.

Simple cases

The Z and Y matrices are inverse one from the other, and the same goes for the H and G matrices. That is,

$$\mathbf{Z}^{-1} = \mathbf{Y}; \quad \mathbf{Y}^{-1} = \mathbf{Z} \tag{10.1a}$$

$$\mathbf{H}^{-1} = \mathbf{G}; \quad \mathbf{G}^{-1} = \mathbf{H} \tag{10.1b}$$

Hence, these conversions become trivial and will not be further considered.

The chain parameters are not inverse because of the negative signs for the currents, but these can be worked with the matrix element operation dot-star of the TI-nspire. The conversion is done as

$$\begin{bmatrix} 1 & -1 \\ -1 & 1 \end{bmatrix} .\times \mathbf{A}^{-1} = \mathbf{B}; \quad \begin{bmatrix} 1 & -1 \\ -1 & 1 \end{bmatrix} .\times \mathbf{B}^{-1} = \mathbf{A} \tag{10.1c}$$

Although the four transformations (10.1a) and (10.1b) are really not worth programming, the two shown in (10.1c) may be programmed as functions in the TI-89 taking advantage of the dot operations:

A to B transformation:
Define a2b(a)= [1, (-) 1 ; (-) 1 , 1] .* a ^ (-) 1

B to A transformation:
Define b2a(b)= [1, (-) 1 ; (-) 1 , 1] .* b ^ (-) 1

Conversion by setting up and solving the equations

The basic principle for conversion by this method is to use the original set of equations and rearrange them in the form $\mathbf{Px} = \mathbf{Qy}$, where $\mathbf{x}$ is the vector of the new dependent variables, and $\mathbf{y}$ that of the new independent variables.

10.2. TWO PORT PARAMETERS

The 2x2 matrices **P** and **Q** are built by rearrangement of coefficients. Let us explain it using a specific case, conversion from Y to H parameters.

We assume that matrix **Y** of Y-parameters is given:

$$\mathbf{Y} = \begin{bmatrix} y_{11} & y_{12} \\ y_{21} & y_{22} \end{bmatrix}$$

To understand the logic of the process, let us state our problem in terms of equations. We are given the set

$$\begin{aligned} I_1 &= y_{11} V_1 + y_{12} V_2 \\ I_2 &= y_{21} V_1 + y_{22} V_2 \end{aligned}$$

and we want to arrive at a system in the form

$$\begin{aligned} V_1 &= h_{11} I_1 + h_{12} V_2 \\ I_2 &= h_{21} I_1 + h_{22} V_2 \end{aligned}$$

Hence, we rearrange the original set of equations that we want to convert so that the dependent variables V_1 and I_2, and independent ones, I_1 and V_2, are identified. That is

$$\begin{aligned} -y_{11} V_1 &= -I_1 + y_{12} V_2 \\ y_{21} V_1 - I_2 &= 0 I_1 - y_{22} V_2 \end{aligned}$$

In matrix form,

$$\begin{bmatrix} -y_{11} & 0 \\ y_{21} & -1 \end{bmatrix} \begin{bmatrix} V_1 \\ I_2 \end{bmatrix} = \begin{bmatrix} -1 & y_{12} \\ 0 & -y_{22} \end{bmatrix} \begin{bmatrix} I_1 \\ V_2 \end{bmatrix} \quad (10.2)$$

and then solve. The following sequence will generate the h-parameter matrix **H** that results after conversion.

Let us now proceed. Use [≡] and select the 2x2 matrix template. Then hit

[(-)] y [1,1] [tab] 0 [tab] y[2,1] [tab] [(-)] 1 [tab] [sto→] p [enter]

to get

$$\begin{bmatrix} -y[1,1] & 0 \\ y[2,1] & -1 \end{bmatrix} \boxed{\text{sto→}} P \quad (10.3)$$

If you do not want to use the template key, the same result can be entered as

[[[(-)] y[1,1],0] [[(-)] y[2,1], 1]] [sto→] P [ENTER]

Continuing with our problem, define

196 CHAPTER 10. TWO-PORT AND THREE-TERMINAL NETWORKS

$$\begin{bmatrix} -1 & y[1,2] \\ 0 & \boxed{(-)}\,y[2,2] \end{bmatrix} \boxed{\text{sto}\rightarrow}\, Q \qquad (10.4)$$

and solve to get the H matrix

$$P^{-1} \boxed{\times}\, Q \boxed{\text{sto}\rightarrow}\, h \boxed{\text{ENTER}} \qquad (10.5)$$

Example 10.1 *Starting with y-parameter matrix* **Y**.

$$\mathbf{Y} = \begin{bmatrix} 232.4E-3 & -12.5E-3 \\ -3.5E-3 & 121.0E-3 \end{bmatrix}$$

the above steps (10.3) *to* (10.5) *results in the target matrix*

$$\mathbf{H} = \begin{bmatrix} 4.303 & 0.067 \\ -0.015 & 0.121 \end{bmatrix}$$

Programming transformations

Tables 10.2, 10.3 and 10.4 shows formulas for transformations. You may check these formulas as an exercise. To do it, rewrite the equations so as to solve for the new dependent variables, as illustrated in the previous example. Solving for (10.3) to (10.4) algebraically, to put an example, yields

$$\begin{bmatrix} \dfrac{1}{y_{11}} & \dfrac{-y_{12}}{y_{11}} \\ \dfrac{y_{21}}{y_{11}} & \dfrac{y_{11}y_{22}-y_{12}y_{21}}{y_{11}} \end{bmatrix}$$

This is precisely the matrix defined with the formulas of the second line in table 10.3. We can therefore define the function

$$\texttt{y2h(y) := (1}\boxed{\div}\texttt{y[1,1])*[[1,}\boxed{(-)}\ \texttt{y[1,2]]\ [y[2,1], det(y)]]} \qquad (10.6)$$

As a matter of fact, I used TI-nspire cx cas symbolic capabilities, to solve the equations!

In a similar way, for each transformation a function can defined with the name shown in the tables. Naturally, you may change the name or not include a function, as you wish.

Observe that all transformations involve dividing either by a determinant or by one parameter. When the division is by zero, the calculator will warn about the error. This means that the target set of parameters does not exist. Remember, however, that due to rounding errors, sometimes the division is not by zero but by a very small number, yielding unreasonable results. Exercise your judgement!

10.3. APPLYING TWO-PORT PARAMETERS

Table 10.2: Two-port parameters Z and Y conversion formulas

$$z_{11} = \frac{\Delta h}{h_{22}}; \quad z_{12} = \frac{h_{12}}{h_{22}}$$
$$z_{21} = -\frac{h_{21}}{h_{22}}; \quad z_{22} = \frac{1}{h_{22}}$$

Define h2z(h)

$$y_{11} = \frac{1}{h_{11}} \quad y_{12} = -\frac{h_{12}}{h_{11}}$$
$$y_{21} = \frac{h_{21}}{h_{11}} \quad y_{22} = -\frac{\Delta h}{h_{11}}$$

Define h2y(h)

$$z_{11} = \frac{1}{g_{11}}; \quad z_{12} = -\frac{g_{12}}{g_{11}}$$
$$z_{21} = \frac{g_{21}}{g_{11}}; \quad z_{22} = \frac{\Delta g}{g_{11}}$$

Define g2z(g)

$$y_{11} = \frac{\Delta g}{g_{22}} \quad y_{12} = \frac{g_{12}}{g_{22}}$$
$$y_{21} = -\frac{g_{21}}{g_{22}} \quad y_{22} = -\frac{1}{g_{22}}$$

Define g2y(g)

$$z_{11} = -\frac{a_{11}}{a_{21}}; \quad z_{12} = \frac{\Delta a}{a_{21}}$$
$$z_{21} = \frac{1}{a_{21}}; \quad z_{22} = -\frac{a_{22}}{a_{21}}$$

Define a2z(a)

$$y_{11} = \frac{a_{22}}{a_{12}}; \quad y_{12} = -\frac{\Delta a}{a_{12}}$$
$$y_{21} = 1\frac{1}{a_{12}}; \quad y_{22} = \frac{a_{11}}{a_{12}}$$

Define a2y(a)

$$z_{11} = \frac{b_{22}}{b_{21}}; \quad z_{12} = \frac{1}{b_{21}}$$
$$z_{21} = \frac{\Delta b}{b_{21}}; \quad z_{22} = \frac{b_{11}}{b_{21}}$$

Define b2z(b)

$$y_{11} = \frac{b_{11}}{b_{12}}; \quad y_{12} = -\frac{1}{b_{12}}$$
$$y_{21} = -\frac{\Delta b}{b_{12}}; \quad y_{22} = \frac{b_{22}}{b_{12}}$$

Define b2y(b)

10.3 Applying two-port parameters

Once you have a set of parameters, you do not need to insert the full sub-circuit, but only work with it using the equations, or at most, the equivalent

Table 10.3: Two-port parameters H and G conversion formulas

$$h_{11} = \frac{\Delta z}{z_{22}}; \quad h_{12} = \frac{z_{12}}{z_{22}}$$

$$h_{21} = -\frac{z_{21}}{z_{22}}; \quad h_{22} = \frac{1}{z_{22}}$$

Define z2h(z)

$$g_{11} = \frac{1}{z_{11}} \quad g_{12} = -\frac{z_{12}}{z_{11}}$$

$$g_{21} = \frac{h_{21}}{z_{11}} \quad g_{22} = -\frac{\Delta z}{z_{11}}$$

Define z2g(z)

$$h_{11} = \frac{1}{y_{11}}; \quad h_{12} = -\frac{y_{12}}{y_{11}}$$

$$h_{21} = \frac{y_{21}}{y_{11}}; \quad h_{22} = \frac{\Delta y}{y_{11}}$$

Define y2h(y)

$$g_{11} = \frac{\Delta y}{y_{22}} \quad g_{12} = \frac{y_{12}}{y_{22}}$$

$$g_{21} = -\frac{y_{21}}{y_{22}} \quad g_{22} = -\frac{1}{y_{22}}$$

Define y2g(y)

$$h_{11} = \frac{a_{12}}{a_{22}} \quad h_{12} = \frac{\Delta a}{a_{22}}$$

$$h_{21} = \frac{1}{a_{22}} \quad h_{22} = -\frac{a_{21}}{a_{22}}$$

Define a2h(a)

$$g_{11} = \frac{a_{21}}{a_{11}} \quad g_{12} = -\frac{\Delta a}{a_{11}}$$

$$g_{21} = \frac{1}{a_{21}} \quad g_{11} = \frac{a_{12}}{a_{11}}$$

Define a2g(a)

$$h_{11} = \frac{b_{12}}{b_{11}} \quad h_{12} = \frac{1}{b_{11}}$$

$$h_{21} = \frac{\Delta b}{b_{11}} \quad h_{22} = \frac{b_{21}}{b_{11}}$$

Define b2h(b)

$$g_{11} = \frac{b_{21}}{b_{22}} \quad g_{12} = -\frac{1}{b_{22}}$$

$$g_{21} = \frac{\Delta b}{b_{22}} \quad g_{22} = \frac{b_{12}}{b_{22}}$$

Define b2g(b)

10.3. APPLYING TWO-PORT PARAMETERS

Table 10.4: Two-port chain parameters conversion formulas

$a_{11} = \dfrac{z_{11}}{z_{21}} \quad a_{12} = \dfrac{\Delta z}{z_{21}}$ $a_{21} = \dfrac{1}{z_{21}} \quad a_{22} = \dfrac{z_{22}}{z_{21}}$ Define z2a(z)	$b_{11} = \dfrac{z_{22}}{z_{12}} \quad b_{12} = \dfrac{\Delta z}{z_{12}}$ $b_{21} = \dfrac{1}{y_{12}} \quad b_{22} = -\dfrac{z_{11}}{y_{12}}$ Define z2b(z)
$a_{11} = -\dfrac{y_{22}}{y_{21}} \quad a_{12} = -\dfrac{1}{y_{21}}$ $a_{21} = -\dfrac{\Delta y}{y_{21}} \quad a_{22} = -\dfrac{y_{11}}{y_{21}}$ Define y2a(y)	$b_{11} = -\dfrac{y_{11}}{y_{12}} \quad b_{12} = -\dfrac{1}{y_{12}}$ $b_{21} = -\dfrac{\Delta y}{y_{12}} \quad b_{22} = \dfrac{y_{22}}{y_{12}}$ Define y2b(y)
$a_{11} = -\dfrac{\Delta h}{h_{21}} \quad a_{12} = -\dfrac{h_{11}}{h_{21}}$ $a_{21} = -\dfrac{h_{22}}{h_{21}}; \quad a_{22} = -\dfrac{1}{h_{21}}$ Define h2a(h)	$b_{11} = \dfrac{1}{h_{12}} \quad b_{12} = \dfrac{h_{11}}{h_{12}}$ $b_{21} = \dfrac{h_{22}}{h_{12}} \quad b_{22} = \dfrac{\Delta h}{h_{12}}$ Define h2b(h)
$a_{11} = \dfrac{1}{g_{21}} \quad a_{12} = \dfrac{g_{22}}{g_{21}}$ $a_{21} = \dfrac{g_{11}}{g_{21}} \quad a_{22} = \dfrac{\Delta g}{g_{21}}$ Define g2a(g)	$b_{11} = -\dfrac{\Delta g}{g_{12}} \quad b_{12} = -\dfrac{g_{22}}{g_{12}}$ $b_{21} = -\dfrac{g_{11}}{g_{12}} \quad b_{22} = -\dfrac{1}{g_{12}}$ Define g2b(g)
$a_{11} = -\dfrac{b_{22}}{\Delta b} \quad a_{12} = \dfrac{b_{12}}{\Delta b}$ $a_{21} = -\dfrac{b_{21}}{\Delta b} \quad a_{22} = \dfrac{b_{11}}{\Delta b}$ Define b2a(b)	$b_{11} = \dfrac{a_{22}}{\Delta a} \quad b_{12} = \dfrac{a_{12}}{\Delta a}$ $b_{21} = \dfrac{a_{21}}{\Delta a} \quad b_{22} = \dfrac{a_{11}}{\Delta a}$ Define a2b(a)

macro model circuit. This one exists for parameters **Z, Y, H**, and **G** (See Fig. 10.2). When you substitute the two-port by its model, then the circuit becomes another one with only two terminals, and may be analyzed using any of the methods presented in previous chapters. There are also many other useful applications which fall beyond our scope.

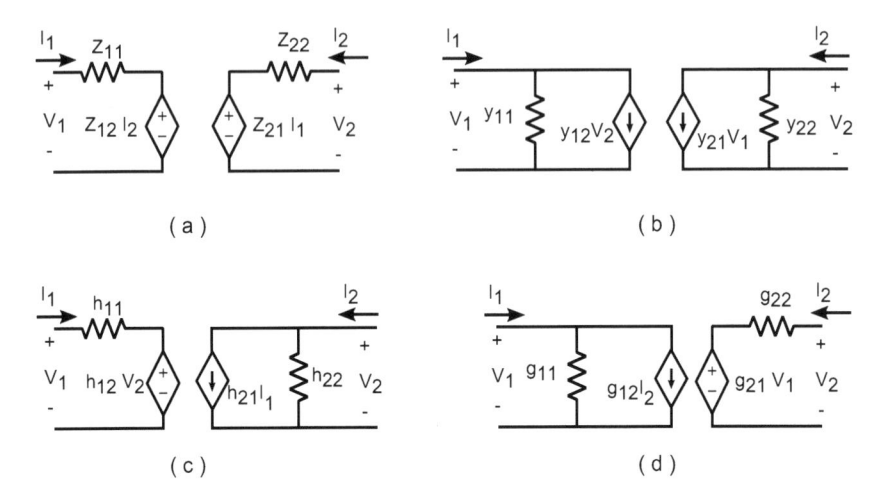

Figure 10.2: Two port equivalent models:(a) from Z-parameters, (b) from Y Parameters, (c) From H parameters, (d) from G parameters

In this section, only two among the many cases are considered: terminated two-ports and, tandem connection.

10.3.1 Terminated two-ports

Fig. 10.3 shows two-port networks terminated by a load resistance and a source together with a resistance Rs, which may represent the Thevenin or Norton equivalent. For these configurations, we can establish formulas for the different network functions which we may program in our calculator using the two port parameters. There is an example below about finding the formulas. The reader may work any formulas he/she wishes, or else look at tables available in textbooks or internet, and program results. I omit here those tables.

Example 10.2 *Assume the two port is characterized by the hybrid parameters:*

$$V_1 = h_{11}I_1 + h_{12}V_2$$

and

$$I_2 = h_{21}I_1 + h_{22}V_2$$

10.3. APPLYING TWO-PORT PARAMETERS

(a)

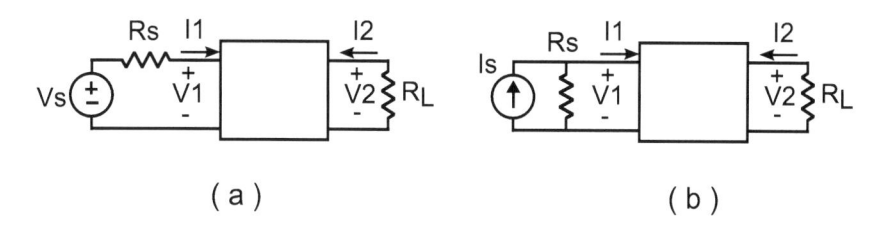

(b)

Figure 10.3: Terminated two ports :(a) with voltage source, (b) with current source

Now, let us consider the additional equations that arise from the connection shown in Fig. 10.3:

$$V_2 = -R_L I_2 \quad and \quad V_1 + R_s I_1 = V_s$$

Using the four equations, we can deduce after some Algebra, or using your calculator symbolic algebra capabilities, the following formulas. Notice that if you plan to use your calculator, then you must rearrange your equations so that a proper parameter stands as a known one and others as unknown ones. It is worth trying this exercise!

$$R_{in} = \frac{V_1}{I_1} = \frac{h_{11} + \Delta h \, R_L}{1 + h_{22} \, R_L}$$

$$I_2 = \frac{h_{21} V_s}{(h_{11} + R_s)(1 + h_{22} R_L) - h_{12} h_{21} R_L}$$

$$R_{th} = \frac{R_s + h_{11}}{\Delta h + h_{22} \, R_s}$$

$$V_{th} = \frac{-h_{21} V_s}{\Delta h + h_{22} \, R_s}$$

$$A_{vs} = \frac{V_2}{V_s} = \frac{-h_{21} R_L}{(h_{11} + R_s)(1 + h_{22} R_L) - h_{12} h_{21} R_L}$$

$$A_v = \frac{V_2}{V_1} = \frac{-h_{21} R_L}{h_{11} + \Delta \, R_s}$$

$$A_I = \frac{I_2}{I_1} = \frac{h_{21}}{1 + h_{22} \, R_L}$$

$$G_m = \frac{I_2}{V_1} = \frac{h_{21}}{h_{11} + \Delta h \, R_L}$$

In this set, V_{th} and R_{th} stand for the Thevenin equivalent seen by load

R_L. From these formulas we could perfectly define functions. For example a function for the input resistance above could be the following

```
Define RinH(h,rl)= (h[1,1] + det(h)× rl)/(1 + rl × h[2,2])
```

We could also write a program to generate all or some results. For example, an example of program with four results is as follows:

```
terminatedh(h,rs,rl)
Pgrm
Local P
(1 + h[2,2]*rl) → P
(h[1,1] + det(h)*rl)/P → rin
Disp ''Rin = '' rin
(h[2,1]/P → Ai
Disp ''Ai = '' Ai
(det(h) + rs*h22) → P
(h[1,1] + rs)/P → rth
Disp ''Rth = '' rth
(-h[2,1]/P → vth
Disp ''Vth/Vs = '' vth
EndPgrm
```

As illustrated with the example, you may build up your own data base of formulas in your calculator using the different parameters for the network functions. Another alternative, equally valid, is to work with the equations directly and interpret your results. Consider the following example.

Example 10.3 *Assume that in Fig. 10.3(b) Rs = 2.12 kΩ and RL = 1.5 kΩ. Furthermore, assume that the two port is described by the reverse chain parameters as*

$$V_1 = 0.125 V_2 - 8 I_2$$

and

$$I_1 = 0.001 V_2 - 1.2 I_2$$

The connection equations are $I_1 + V_1/R_s = I_s$ and $V_2 = -R_L I_2$. Combining all equations with the given values, and taking $I_s = 1$ A for network function calculation, we arrive at:

$$\begin{array}{c} \\ V_1 \\ I_1 \\ Input \\ Out \end{array} \begin{array}{cccc} V_1 & V_2 & I_1 & I_2 \\ \left[\begin{array}{cccc} -1 & 0.125 & 0 & -8 \\ 0 & .001 & -1 & -1.2 \\ \frac{1}{2120} & 0 & 1 & 0 \\ 0 & 1 & 0 & 1500 \end{array} \right] \end{array} \left[\begin{array}{c} V_1 \\ V_2 \\ I_1 \\ I_2 \end{array} \right] = \begin{array}{c} Is \\ \left[\begin{array}{c} 0 \\ 0 \\ 1 \\ 0 \end{array} \right] \end{array} \quad (10.7)$$

10.3. APPLYING TWO-PORT PARAMETERS

Solving this equation we arrive at

$$\begin{bmatrix} V_1 \\ V_2 \\ I_1 \\ I_2 \end{bmatrix} = \begin{bmatrix} 70.016 \\ 537.208 \\ 966.97E-3 \\ -358.14E-3 \end{bmatrix} \tag{10.8}$$

Let us store this result in variable q. From the vector we can deduce the following:

$\dfrac{V_2}{I_s} = 537.8\ \Omega(\ \text{q[2,1]}),\quad \dfrac{-I_2}{I_s} = 0.358\ (\ \boxed{(-)}\ \text{q[4,1]})\quad R_{in} = \dfrac{V_1}{I_s} =$
q[1,1] = 70.016 Ω

and so on.

10.3.2 Tandem Connection

A particular structure that in general is simpler to work with two parameters, and easy to program, is the one known as a *cascade* or *tandem* connection of two ports, illustrated in Fig. 10.4. Here, the "output" port 2 of block 1 is connected to the "input" port 1 of the next two-port. The overall system may be considered as a two-port, in which port 1 is the port of the first two-port in the chain, while port 2 is that of the last one in the chain. Notice that $V_{21} = V_{12}$ and $-I_{21} = I_{12}$, which explains why we use the negative of current I_2 in the definition of chain parameters.

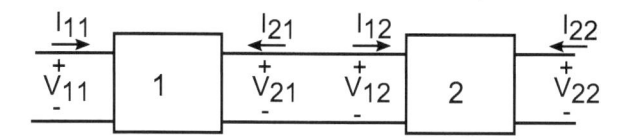

Figure 10.4: Cascade or tandem connection of two ports.

Looking at Fig. 10.4, we see that

$$\begin{bmatrix} V_{11} \\ I_{11} \end{bmatrix} = \mathbf{A_1} \begin{bmatrix} V_{21} \\ -I_{21} \end{bmatrix} = \mathbf{A_1} \begin{bmatrix} V_{12} \\ I_{12} \end{bmatrix} = \mathbf{A_1}\,\mathbf{A_2} \begin{bmatrix} V_{12} \\ -I_{22} \end{bmatrix}$$

This means that we can obtain the overall chain parameters using matrix multiplication of the individual port matrices, in the same order in which they are connected.

This principle is applicable in many cases of interest including programming of configurations where two-port configurations in cascade may be identified. Among these configurations we find the popular ladder network, shown in Fig. 10.5. In this circuit, the cascade connection is identified as a successive tandem placement of the two simple ports called series resistance and parallel resistance, which are shown in Fig. 10.6(a) and (b).

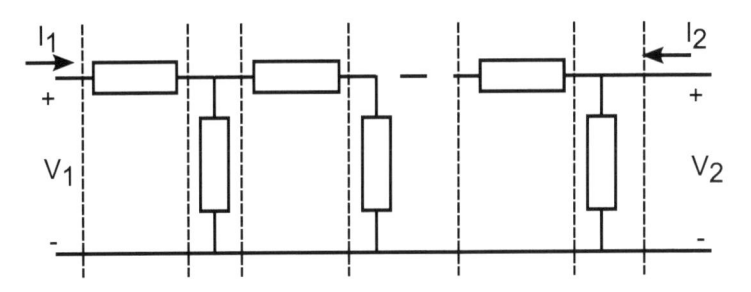

Figure 10.5: Ladder configuration.

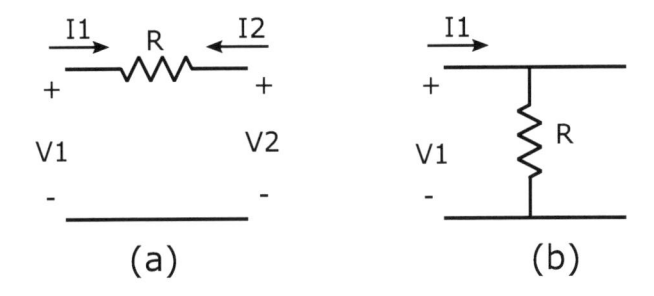

Figure 10.6: Examples of two ports for data base: a) Series R, b) Parallel R

For these two cases, the chain matrices for the series and parallel resistance two ports are, respectively

$$\begin{bmatrix} 1 & R \\ 0 & 1 \end{bmatrix} \quad (10.9)$$

and

$$\begin{bmatrix} 1 & 0 \\ \frac{1}{R} & 1 \end{bmatrix} \quad (10.10)$$

There are of course other examples of two ports which you can connect in tandem. Two examples are those in in Fig. 10.7, an ideal voltage controlled current source and a voltage amplifier with infinite input resistances. These structures have the following matrices, respectively.

$$\begin{bmatrix} 0 & -\frac{1}{g} \\ 0 & 0 \end{bmatrix} \quad (10.11)$$

10.3. APPLYING TWO-PORT PARAMETERS

$$\frac{1}{G_f R_o + A} \begin{bmatrix} G_f R_o + 1 & R_o \\ G_f(1-A) & G_f R_o \end{bmatrix} \qquad (10.12)$$

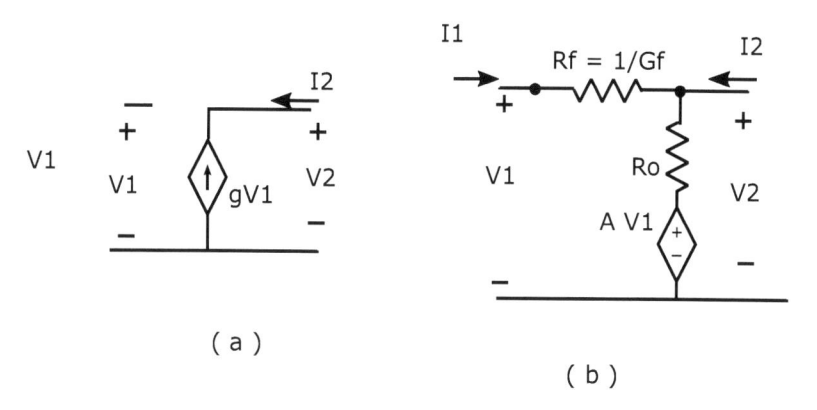

Figure 10.7: Other examples of two ports for data base: a) ideal VCCS, b) Voltage amplifier with infinite imput resistance

Notice that the conductance G_f form of the resistance is used. This is because an open circuit has conductance 0, and therefore can be used in programming.

All four cases may be programmed as functions, for example, or else included as subroutines in a program. You can program it using the matrix editor directly. For example

$$\texttt{serr(r)} := \begin{bmatrix} 1 & r \\ 0 & 1 \end{bmatrix}$$

Or you can use an entry on the command line. Examples are:

Series R branch: `Define serr(r)= [ 1, r ; 0 , 1 ]`

Parallel R branch: `Define parr(r)= [ 1, 0 ; 1 / r , 1 ]`

VCCS: `Define vcs2(g)= [ 0 , (-) 1 / g ; 0 , 0 ]`

Rf_Ro_A two port: (Ro =r, 1/Rf = gf with gf=0 when Rf = ∞)

`rga(r,gf,a):=(1/(gf`×`r+a))`×`[(gf`×`r+1), r; gf`×`(1 - a),gf`×`r]`

Notice that an ideal voltage controlled voltage source is included for the case `rga(0,0,a)`.

206CHAPTER 10. TWO-PORT AND THREE-TERMINAL NETWORKS

Thus, you can create your own database with the two-ports that you use frequently. An example using these functions for the tandem connection follows.

Example 10.4 *Consider the loaded inverting amplifier of Fig. 10.8(a), which has an ideal voltage gain $V_o/V_s = -10$. Assuming an operational amplifier with an input resistance of 200 kΩ, an output resitance of 75 Ω and an open circuit gain of 1000, the equivalent circuit is that of Fig. 10.8(a). Find the actual gain.*

(a)

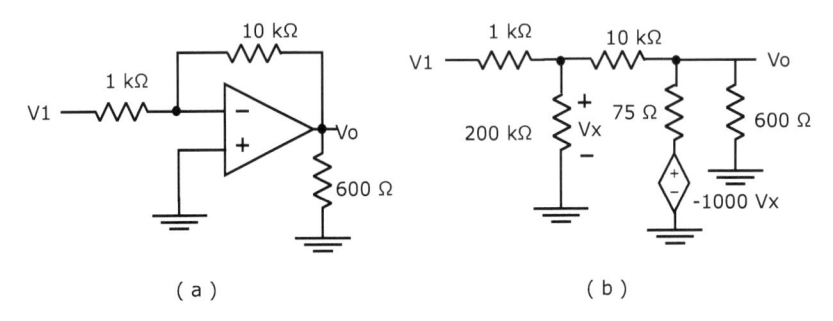

(b)

Figure 10.8: a) A loaded inverting amplifier, and b) its equivalent circuit.

We can identify the two ports in cascade for this configuration, considering again the last port as a one terminated with an open circuit. Using then the functions defined before we get the overall chain matrix with the input

serr(1000)*parr(200E3)*rga(75,1/10000, (-) 1000)*parr(600) ENTER
to get

$$\begin{bmatrix} -101.25\text{E-}3 & -82.88\text{E-}3 \\ -100.12\text{E-}6 & -7.88\text{E-}6 \end{bmatrix}$$

From this result, the actual gain is obtained with

$$1/\text{ans}[1,1] \rightarrow -9.8764$$

Consider now another example. This one allows us to look at a terminated two port network from another perspective.

Example 10.5 *Take again the terminated two-port of example 10.3 which is redrawn in Fig. 10.9 so that the cascade structure is identified.*

We should be aware of the way the procedure works: 1) I_2 of the two-port *is not* the current in the 1.5 kΩ resistor. For the combined two-port, $I_2 = 0$; (b) the current in the resistor is $I_L = V_2/1500$; and (c) for the combined two-port $I_1 = I_s = 1$.
We can now apply the tandem property:

10.4. CLOSING REMARKS

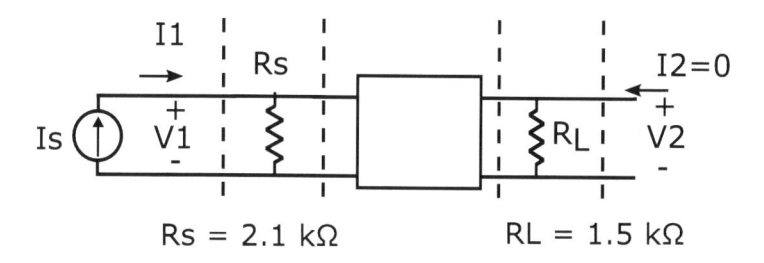

```
          I1        Rs                              I2=0
         →     +                              +
    Is ( ↑ ) V1     ⌇                  ⌇ RL    V2
               −                              −

              Rs = 2.1 kΩ              RL = 1.5 kΩ
```

Figure 10.9: Terminated two-port interpreted as cascade of two-ports.

$$\begin{bmatrix} A & B \\ C & D \end{bmatrix} = \begin{bmatrix} 1 & 0 \\ \frac{1}{2120} & 1 \end{bmatrix} \begin{bmatrix} 0.125 & 8 \\ .001 & 1.2 \end{bmatrix} \begin{bmatrix} 1 & 0 \\ \frac{1}{1500} & 1 \end{bmatrix}$$

$$= \begin{bmatrix} 130.330\text{E-}3 & 1.204\text{E}0 \\ 1.861\text{E-}3 & 1.204\text{E}0 \end{bmatrix}$$

Storing again the new matrix in variable q, *we get*

$$\frac{V_2}{I_s} = 537.8 \text{ } \Omega (1 \text{ / } \text{q[2,1]}), \quad \frac{I_L}{I_s} = 0.358 \text{ } (1/ \text{ q[2,1]}/1500)$$

$$R_{in} = \frac{V_1}{I_1} = \frac{\text{q[1,1]}}{\text{q[2,1]}} = 70.016 \text{ } \Omega$$

We see that with the proper interpretation we obtain the same results as in example 10.3, but with less effort.

10.4 Closing remarks

The topic of two ports can be extended. You can find more formulas which can be added to your data base of user defined function.

Further topics could include on how to insert three terminal networks when embedded in a circuit ([Moschytz1974]), multiports, cases with independent sources, s-parameters, and others. Some will be considered in the companion book. But the important thing is that you have the tools to continue by yourself if interested. Remember, the calculator is your tool, you are the learner.

CHAPTER 11

What's Next

Well, I must finish somewhere. I decided to limit this introductory book to resistive circuits only, and this is why it has the subtitle "First part, resistive circuits" . Obviously, you may object that there are still too many topics not covered. And you're right. But I certainly covered enough to be sure that you can go ahead by yourself in many ways. This chapter is some sort of an appetizer for going further into circuit analysis using the calculator, introducing some topics left for the companion book, and also letting you know some extra topics you can already explore with the knowledge acquired with this book.

Two brief fields are explored as an introduction and also as a showcase that we have advanced more than you might expect. One is a visit to phasor domain circuits, pertaining to steady state analysis of reactive circuits with sine signals. The other is an example of first order linear circuits.

Dealing with linear circuits containing reactive elements is also another topic to be expanded in a later book. In section 11.2, I show with an example that this area is already advanced with the methods already developed here.

Don't forget, however, that the calculator is not the teacher, only the tool.

11.1 Complex circuits in steady state domain

Circuits containing resistive, inductive and capacitive elements, as well as linear dependent sources which may be also frequency dependent, are dealt with using complex numbers in the phasor domain. Two different venues should be considered. One, is that of applications and of theoretical nature. The other concerns the analysis of circuits, that is, obtaining the current and voltage phasors. While the first one clearly falls outside the scope of this book and will be considered in part in the companion volume, the second is

11.1. COMPLEX CIRCUITS IN STEADY STATE DOMAIN

already "covered", to say it somehow.

This is real. All the analyses done here with purely resistive elements are directly applicable to circuits with complex impedances and admittances, with the sources being constant complex numbers, namely, phasors. The TI-nspire cx cas allows handling of complex numbers in lists and matrices, so you should find no problem in applying the methods and programs mentioned here.

If all independent sources are in the form $K\cos(\omega t + \phi)$, and all of the same angular frequency ω, then the source is associated to a complex number $K\angle\phi$ (explained next). All complex currents and voltages in the circuit will be complex numbers which can be then associated to a cosine function of frequency ω.

The theory behind this process is found in your textbook, so I don't work with it anymore in this book. Let me only show how your knowledge is applied.

11.1.1 Elementary notation for complex numbers

A complex number z is in rectangular form if it is written as $z = a + bi$, where $i = \sqrt{-1}$ is the imaginary unit. The real and imaginary parts of the complex number z are defined as $\text{Re}(z) = a$, $\text{Im}(z) = b$, respectively. **In circuits we use j for the imaginary unit, but calculators use i.** In this book, we adhere the the j notation for the circuits but use i for the calculator keys.

On the other hand, in polar form, z is written as $z = r\angle\phi = re^{i\phi}$. Here, the absolute value of z is $\text{abs}(z) = |z| = r$ and the angle or argument of z is $\arg(z) = \text{angle}(z) = \phi$. We always assume $r \geq 0$. $\angle\phi$ is the shorthand notation for the Euler identity

$$e^{i\phi} = \cos\phi + i\sin\phi \qquad (11.1)$$

usually used in engineering textbooks.

Graphically, the complex number z may be represented by a vector in a two dimensional Cartesian plane, called complex plane. The real part is represented in the x-axis, and the imaginary part in the y-axis. The absolute and angle values are then the polar coordinates of the vector. Very often in circuits we try to limit the angle to the values $-180° \leq \phi \leq 180°$. Fig. 11.1 illustrates some examples for complex numbers.

11.1.2 Complex numbers and calculator

To work with complex numbers in your calculator prepare your settings accordingly. Do not work in Real mode, because your calculator returns an error for a complex-number result, except if you enter your expression using i, the imaginary unit.

To enter the imaginary unit **i***: press* $\boxed{\pi}$ $\boxed{\pi\ \texttt{enter}}$

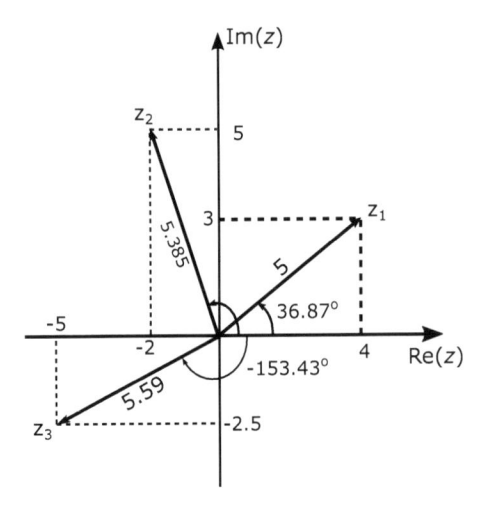

Figure 11.1: Graphical representation of $z_1 = 4 + 3i = 5\angle 36.87°$, $z_2 = -2 + 5i = 5.385\angle 111.80°$ and $z_3 = -5 - 2i = 5.59\angle -153.43°$ in the complex plane.

Settings and Display

For your settings, you must select two items:

Real or Complex Select either rectangular mode $(a+bi)$ or complex polar mode

Angle Must choose angle settings (radians, degrees, gradians) for polar mode display and input

When in degrees or gradian mode, the output display is in the form $(r\angle \phi)$. For example

In degree mode: 3 + i → (3.1623 ∠ 18.435)

In gradian mode: 3 + i → (3.1623 ∠ 20.483)

On the other hand, if the angle mode is in radians, then the output is in the form $e^{\phi i}$ r. For example

In radian mode: 3 + i → $e^{.32175i}$ 3.1623

11.1. COMPLEX CIRCUITS IN STEADY STATE DOMAIN

Angle setting and polar input

The input can be in any mode, rectangular or polar. The real or complex mode sets the display of results. Complex input specifying absolute value and angle must be enclosed in parenthesis. The angle ($\angle$) entry is available in the $\infty\beta^o$ menu. The display of the result depends on angle settings. For example:

In degree mode: ($\sqrt{2}\angle 45$) $\to$ 1. + i

In radian mode: ($\sqrt{2}\angle 45$) $\to$ 0.74292 + 1.2034 i

In gradian mode: ($\sqrt{2}\angle 45$) $\to$ 1.0754 + 0.91846 i

Polar inputs in exponential form are valid only for radian mode.

In radian mode: $\sqrt{2}e^{45i}$ $\to$ 0.74292 + 1.2034 i

In degree mode: $\sqrt{2}e^{45i}$ $\to$ ``Domain error''

Angle settings can be overridden with the inputs ^{o} to enter degrees, ^{r} to enter radians and ^{g} to enter gradians. These entries are available in the $\boxed{\pi}$ key. Thus, in any angle mode

($\sqrt{2}\angle 45^o$) $\to$ 1. + i

$\sqrt{2}e^{45^o i}$ $\to$ 1. + i

11.1.3 An Example

Let us see one example using transformations and applying lists. The solution of the problem could also be done using nodal or loop equations I encourage the reader to try them.

Example 11.1 *Consider example 5.20 on page 84. In this example, the current I was found using transformations and lists. Let us now work a "similar" problem, with the same topology, asking for a current in the same branch, and following the same steps, but in the complex plane.*

Consider the circuit in Fig. 11.2(a); the solution was obtained then reducing the circuit to that shown in (b) by first applying source translation and then Thevenin equivalent calculated simultaneously for both extremes. I recommend the reader to see again the original example where the steps are illustrated graphically. Then, draw the steps for this examle.

Now, I will follow exactly the same steps as in example 5.20, except for the fact that I am using complex numbers in the lists, and for the voltage. Settings for the display are polar and degrees for the display, with two decimal

212   CHAPTER 11. WHAT'S NEXT

figures, **except for current where ENG mode is used** *since the current magnitude in mA is not displayed with the original settings. Notice that polar inputs are in parenthesis.*

(a)

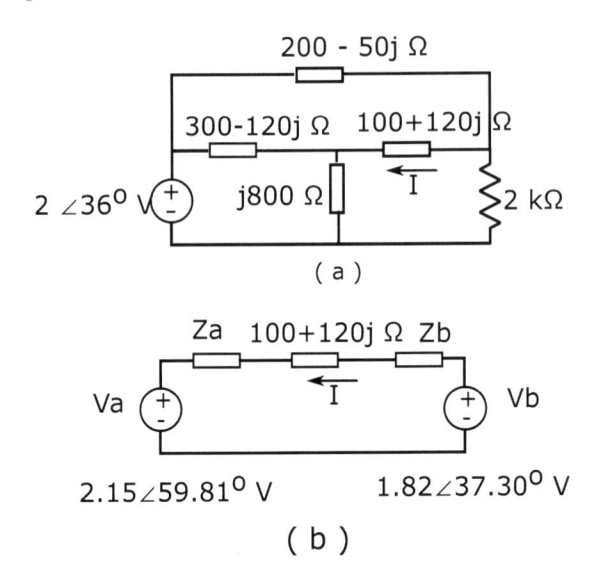

(b)

Figure 11.2: Reduction for a complex bridged-T. Za = $(347.79\angle 2.00)$ Ω, Zb = $(187.37\angle -12.73)$ Ω

The table that follows is similar to the one used in the original example 5.20.

Entry: {300-120i, 200-50i} [sto→] Z1
Result in Stack: {(323.11∠ -21.80) (206.16∠-14.04)}

Entry: {800i, 2000} [sto→] Z2
In Stack {(800.∠ 90.)   2000.}

Entry: (2∠36) [×] Z2 [÷] (Z1 + Z2) [sto→] Vt
Result in Stack: {(2.15∠59.81) (1.82∠37.30)}
Meaning: Vt = {Va, Vb}

Z1 [×] Z2 [÷] (Z1 + Z2) [sto→] Zt
Result in Stack {(347.79∠2.00) (187.37∠ -12.73)}
Meaning: {Zta, Ztb}

(Vt[2]-Vt[1]]) [÷] (100 + 120i + sum(Zt))
Result in Stack (1.32E-3∠-72.63)
Meaning: Current sought

11.2. SHORT REFERENCE TO TIME DOMAIN CIRCUITS

The complex current that results is hence $1.32\angle -72.63°$ mA. In rectangular form, you may convert to get $(0.394 - 1.26\,i)$ mA. Remember though that voltages and currents are usually preferred in polar form because of its relation with the time domain expression.

The reader can verify that the user defined functions introduced, as well as any other that the reader has defined, are valid with complex numbers. The same goes with the programs nodal1, mesh1 and so on. In other words, as far as the tools needed, all the previous chapters are valid in the complex domain.

11.2 Short reference to time domain circuits

To the companion book I have also left the introductory topic to time domain analysis. As an appetizer, consider the conceptual example shown in Fig. 11.3 for a first order circuit. The case for an inductor can also be considered. Inset (a) shows the general configuration. To find the current and voltage at the capacitance C, we calculate the Thevenin equivalent, for example using a current source (Section §9.1), as seen in inset (b). We arrive then at the simple RC circuit of Fig. 11.3(c), in which we find voltage $V_c(t)$ and current $i_c(t)$.

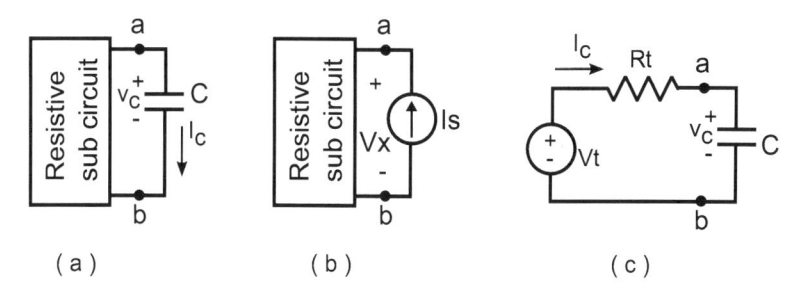

Figure 11.3: A first RC circuit: (a) Original configuration; (b) Ready to find the equivalent seen by capacitor; (c) Circuit to find current and voltage at capacitor.

Once you have the current $i_c(t)$, the substitution theorem (Section 4.1) states that the capacitor may be substitute by a current source, which is the case of (b) if you make $I_s = -i_c(t)$. This means that you already have all the information, because of superposition theorem, to find out all the voltages and currents in your circuit if you desire to do so. Let us work one final example.

Example 11.2 *Let us take the circuit from example 9.1 on page 177, terminated with a capacitance of 12 µF as shown in Fig. 11.4. Let us assume that the initial condition at the capacitance is $V_C(0) = 1$ V*

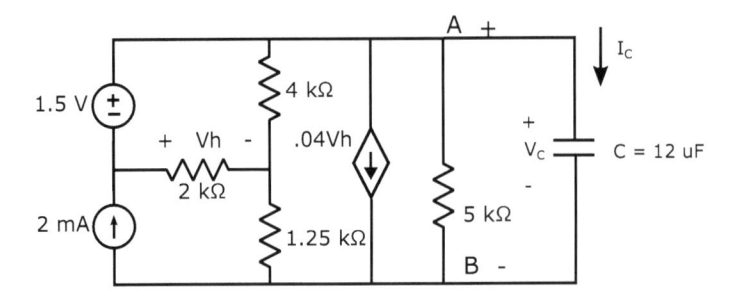

Figure 11.4: A first order circuit to illustrate the procedure

Since the sources in the circuit are constant, the theory of first order circuits tells us that the voltage $V_C(t)$ and current $I_C(t)$ at the capacitance for $t \geq 0$ are given by

$$V_C(t) = V_{th} + (V_C(0) - V_{th})e^{-t/\tau} \qquad (11.2)$$

and

$$I_C(t) = \frac{V_{th} - V_C(0)}{R_{th}} e^{-t/\tau} \qquad (11.3)$$

where we have the time constant

$$\tau = R_{th} C \qquad (11.4)$$

Substituting the capacitance by a current source, as it is illustrated by Fig. 11.3(b), takes us to Fig. 9.5(a) on page 178. For this circuit, the nodal equations were set in the example and solved. Let us store the solution in variable **B**. *From the example we have*

$$\text{yn}^{-1} * \text{in} \rightarrow \begin{bmatrix} 526.74\text{E-}3 & 47.098\text{E}0 \\ 496.818\text{E-}3 & 22.789\text{E}0 \\ 2.0267\text{E}0 & 47.098\text{E}0 \end{bmatrix} \boxed{\text{sto}\rightarrow} \text{ B} \qquad (11.5)$$

From this result we get as a first interpretation:

b[3,1] = Vth = 2.03 V and b[3,2] = Rth = 47.098 Ω.
From these interpretation we apply (11.2) to (11.4):
τ : b[3,2] × 12 $\boxed{\text{EE}}$ $\boxed{(-)}$ 6 → 565.18E-6
1/τ: $\boxed{\wedge}$ $\boxed{(-)}$ 1 → 1.7694E3
$V_C(0) - V_{th}$: 1 - b[3,1] → -1.0267
Coefficient in I_C : $\boxed{(-)}$ ANS/b[3,2] $\boxed{\text{sto}\rightarrow}$ c → 21.8E-3

Therefore, for $t \geq 0$,

$$V_C(t) = 2.0267 - 1.0267e^{-1769.4t} \quad \text{V}$$

and

$$I_C(t) = 21.8e^{-1769.4t} \quad \text{mA}$$

Furthermore, from superposition (11.5) provides us the expression for all node potentials, since source $Ix = -I_C$. For example, from the first line of the solution, $V_1 = 0.527 - 47.098 I_C$. We can perform the following operation to find all results:

$$\mathbf{B} \boxed{\times} \begin{bmatrix} 1 & 0 \\ 0 & -c \end{bmatrix} = \begin{bmatrix} 526.74E-3 & -1.0267 \\ 486.81E-3 & -496.81E-3 \\ 2.0267 & -1.0267 \end{bmatrix}$$

Observe that the third row are the coefficients for V_C, as expected. The first row tells us that $V_1 = 0.527 - 1.027e^{-1769.4t}$, and similarly for the other row.

Once we have the potentials, we can work any other variable such as current or power.

11.3 Closing remarks

As the examples above show, with the methods presented in this book you already have the fundamentals to continue with more applications. Of course, there are so many more things and methods to discover. So, the answer to "what's next?" might be, "you may go on, you have the tools now."

But never forget: the calculator is your tool, not your teacher. It's up to you to make good use of it.

CHAPTER 12

References

The references below provide some sources I have used either to present the theoretical methods, learn about calculators for tricks that are not included in the guidebooks, and fundamentals to justify the methods presented here for using the calculator. Although some of the techniques presented here for exploiting the calculator were independently developed by myself, I cannot claim being the first one using them, since sometimes I discovered later that the method had already been introduced by someone else.

In the mentioned references there are other methods that may be implemented in the calculator. The interested reader may consult a reference to find out more of these, and have fun in addition to learning.

There is a lot of room for improvement and for introduction of more methods. Up to you!

Bibliography

[Ayres1962]   Ayres Jr, Frank, *Matrices*, Schaum Publishing Co., New York, 1962.

[Balabanian69] Balabanian, Norman and Bickart, Theodore A and Seshu, Sundaram, *Electrical network theory*, John Wiley & Sons, Inc, New York, 1969

[Boite1976]   Boite, René, and Neirynck, Jacques. *Thorie des rseaux de Kirchhoff*. Georgi, 1976.

[Budak87]   Budak, Aran, *Circuit Theory Fundamentals and Applications*, Prentice-Hall, Englewood Cliffs, NJ, 1987.

[Chua69]   Chua, Leon O, *Introduction to nonlinear network theory*, McGraw-Hill, New York, 1969.

[Karni1966]	Karni, Shlomo, *Network Theory: Analysis and Synthesis*, Allyn and Bacon, 1966.
[Kiss68]	Kiss, W.F., Gilson, R.A, *On the formulation of the indefinite matrix*, IEEE J. of Solid-State Circuits, vol. 3, No. 3, pp. 307-308, 1968
[McCalla11]	McCalla, J. and Ouellette S. *TI-Nspire for Dummies*, John Wiley & Sons, Inc, New York, 2011
[Moschytz1974]	Moschytz, George, *Linear Integrated Networks: Fundamentals*, Van Nostrand-Reinhold, New York, 1974.
[Palomera06]	Palomera-García, Rogelio. "Multipole and Multiport Analysis", in *Wiley Encyclopedia of Electrical and Electronics Engineering online* (John G. Webster, Ed.), John Wiley & Sons, Inc, (2001, updated 2006), URL: http://dx.doi.org/10.1002/047134608X.W2506
[ti-uk]	https://education.ti.com/sites/UK/downloads/pdf/First_Steps_with_TI-Nspire_CX.pdf
[Vlach83]	Vlach, J., and Singhal, K. *Computer methods for circuit analysis and design.* Springer Science & Business Media, 1983
[Voltmer99]	Voltmer, David R and Yoder, Mark A, *Electrical Engineering Applications with the TI-89*, Texas Instruments, Inc., 1969

Printed in Great Britain
by Amazon